The Sheep People

The Sheep People:
The Ontology of Making Lives, Building Homes and Forging Herds in Early Bronze Age Norway

Kristin Armstrong Oma

Published by Equinox Publishing Ltd.

UK: Office 415, The Workstation, 15 Paternoster Row, Sheffield, South Yorkshire S1 2BX

USA: ISD, 70 Enterprise Drive, Bristol, CT 06010

www.equinoxpub.com

First published 2018

British Library Cataloguing-in-Publication Data

A catalogue record for this book is available from the British Library.

ISBN 9781781792513 (hardback)

Library of Congress Cataloging-in-Publication Data

Names: Oma, Kristin Armstrong, author.
Title: The sheep people: the ontology of making lives, building homes and forging herds in Early Bronze Age Norway / Kristin Armstrong Oma.
Description: Sheffield, UK ; Bristol, CT : Equinox Publishing Ltd, 2018. |
Includes bibliographical references and index. | Description based on print version record and CIP data provided by publisher; resource not viewed.
Identifiers: LCCN 2017004442 (print) | LCCN 2017011969 (ebook) | ISBN 9781781796009 (ePDF) | ISBN 9781781792513 (hardback)
Subjects: LCSH: Bronze age—Norway. | Sheepherding—Norway—History—To 1500. | Shepherds—Norway—History—To 1500. | Human-Animal relationships—Norway—History—To 1500. | Farmhouses—Norway—History—To 1500. | Longhouses—Norway—History—To 1500. | Wool—Norway—History—To 1500. | Norway—Antiquities. | Social archaeology—Norway.
Classification: LCC GN778.22.N8 (ebook) | LCC GN778.22.N8 O45 2017 (print) | DDC 948.1/01—dc23
LC record available at https://lccn.loc.gov/2017004442

Typeset by Queenston Publishing, Hamilton, Ontario, Canada

Printed and bound in Great Britain by Lightning Source Inc. (La Vergne, TN), and Lightning Source UK Ltd. (Milton Keynes)

Contents

List of Figures

List of Tables

Acknowledgements

Ever since I was a little girl I have been enchanted by sheep. In the process of getting closer to the transition from two-aisled to three-aisled houses in the Early Bronze Age in Rogaland, I became re-enchanted. I am thankful to my father, Magnor Erlend Oma (1947–2011), for dragging me around as his assistant farm-hand from the age of 5 or so. He taught me the fundamental lesson that animals are sentient beings, and individuals that should be respected. He also showed me how interesting and instructive it can be to observe animals closely, by giving them time and space to be themselves.

The single most important person in the writing of this book is not one of the sheep people, but a horse woman. I am indebted to my wonderful mentor Lynda Birke who opened her heart and her home to me, let me ride her horses, play with her dog, and lounge in her jacuzzi—and discussed theoretical Human-Animal stuff through it all.

Thanks a million to dear colleagues at the Museum of Archaeology—foremost my flamboyant boss and good friend, Håkan Petersson, for believing in me, supporting me and giving me snatches of space and time to think. I am very grateful to museum director Ole Madsen for giving me a short sabbatical when I most needed it.

Further, I wish to thank my lovely colleagues in the research network ARI (Animals mediating the real and the imaginary), Siv Kristoffersen, Sean Dexter Denham, Gitte Kjeldsen, Trond Meling, Kristine Orestad Sørgård, Inge Særheim, Håkon Reiersen and Arild Klokkervoll. I am deeply grateful for being a member of the research group "the farm as a social arena," which has provided me with invaluable feedback: Liv Helga Dommasnes, Alf Tore Hommedal, Birgitta Berglund, Helge Sørheim and Inger Storli. Thanks also to the Norwegian-Estonian research network "Animals in changing environments: cultural mediation and semiotic analysis" led by Morten Tønnessen and Timo Maran, with participants Laura Kiiroja, Kadri Tüür, Silver Rattasepp, Paul Thibault, Riin Magnus and Nelly Mäekivi.

I am very grateful to Inger Horve, who generously shared her "house book" with me, and helped me with a number of practical things, and never minded me dropping by to discuss Bronze Age houses. Thanks to Ellen Tjørhom Bøe for nonstop enthusiasm for wool and textiles. Thanks also to fellow Bronze Age researchers Sophie Bergerbrant, Hilde Fyllingen, Lisbeth Prøsch-Danielsen, Knut Ivar Austvoll, Lene Melheim,

Christopher Prescott and Katharina Becker for stimulating discussions and important suggestions.

Very special thanks go to the research group on agrarian societies at the Museum of Archaeology for challenging, constructive and assuring feedback: Hilde Fyllingen, Daniel Fredh, Sara Westling, Eli-Christine Soltvedt, Even Bjørdal, Olle Hemdorff and Trond Meling all read parts of the text very closely and gave me invaluable feedback.

Thank you to friends and colleagues who read and commented drafts at different stages of the making of the book: Sophie Bergerbrant, Kristin Sinnes, Anne Jorunn Frøyen and Niall Armstrong.

I am indebted to several gifted photographers and illustrators who kindly allowed me to use their work: Kristin Sinnes, Priscilla Weaver, Terje Tveit, Kjersti Vevatne, Ragnar Børsheim and most of all artist Anne Helen Robberstad for the use of her stellar photographs. I am very grateful to Mari Høgestøl and Bente Brønnick for facilitating access to photographs and figures that are the property of the Museum of Archaeology, University of Stavanger. This book would not be the same if Theo Bell Gil had not redrawn house plans and other figures for me—thank you!!!

I would like to thank Janet Joyce at Equinox Publishing for publishing this volume. I would also like to express my deep-felt gratitude to my editor Dr Russell Adams for his wonderful work on this book and invaluable corrections and suggestions.

My farming roots are continually nourished by my friends and family, especially by my uncle Tarald Oma and my cousin Tarald Magne Oma (whose conversations awakens in me the most important ideas), and also by Anne Helen Robberstad. Thanks also to the sheep people I have met through the years I have worked on this project. I am grateful to David Kennard and Rosamund Young who both graciously shared their insights with me over mugs of builder's tea.

Thanks to my very dear friend Kate Syvertsen for coming with me on Bronze Age rock art field trips, laughter and tears. You rock!

I am grateful to friends and family that have supported me through the ups and downs of this project, particularly Niall, Oisin, Bessie, Poppy and Ronja, my mum and Misha for the interspecies household we together create. Special thanks to my bff Marilyn for always being there for me.

Eternal gratitude to Niall—my true north—for neverending discussions about animals in every kind of manner.

I dedicate these pages to Bessie, who more than anyone taught me what a magnitude a human-animal relationship can hold.

— 1 —

Towards an Archaeology Informed by Human-Animal Studies

Sheep—dog—human: Dynamics of a triad

What was it like to be a human, a sheep or a sheepdog in the Bronze Age? We can never know, but in the tradition of Thomas Nagel's musings (1974) on what it is like to be a bat, we can *imagine*. And this is what I imagine:

> A dog gazes steadily at a small flock of sheep, holding his body low. Beside him is a shepherd, with arms stretched out wide, as if embracing the flock with care and to protect. The dog never looks at the shepherd, but is still acutely aware of his every command. Together dog and shepherd are "lifting"[1] the sheep, working two as one to get the flock to move in the right direction. A complex relationship is enacted: man and sheep, sheep and dog, dog and man. They must all be socialized together in order for this ritual dance to move according to choreography, a nice and easy lift, the flock staying together and the dog stealthily pacing the ground behind them, without nipping the sheep to nudge them along, and without too much swearing from the shepherd.

This scene could describe an ordinary day in the life of shepherd, sheep and sheepdog at any point in time, from the dawn of animal domestication; to the acclaimed BBC TV-show "One man and his dog"; bucolic narratives of herding (e.g. Hutson 1993 [1910]; Kennard 2004, 2005; Rebanks 2016); a sheepdog trial or one of the training manuals on border collies—the sheepdog *par excellence*.

But the scene I wish to evoke is a rock art panel from Valhaug, in the municipality of Sola on Jæren, southwestern Norway (Figure 1.1). The choice of motif renders the panel truly remarkable, and the scene is uncommon for the Bronze Age; a shepherd with his arms open drives a flock of four horned animals. A smaller animal, probably a dog, seems to aid the shepherd in rounding up the animals (Armstrong Oma 2010, 2013c; Høgestøl and Prøsch-Danielsen 2006). I showed a picture of the rock art panel to professional shepherd David Kennard[2] and he immediately recognized the scene as one he has

1. "The lift" is in sheepdog terminology when the sheep begin to move under the influence of the dog. It should be controlled and orderly.
2. David Kennard has been a professional shepherd all his adult life, and has written several books and made documentaries and tv shows about his work with sheep and dogs. Together with his dogs Swift and Greg he won the national championship of sheepdog trials, in the most prestigious "brace" class—

Figure 1.1 The rock art panel from Valhaug, Sola kommune, Rogaland. Photograph by Kristin Armstrong Oma.

engaged in time and again with his flock and his dogs. The stance of the dog as well as the positioning of shepherd and dog in relation to the flock is peculiar to shepherding.

The motif has a timeless quality—we have ample documentation from modern and historical times of how indispensable sheepdogs are to shepherds, and how one dog can do the work of ten men (beautifully illustrated by Kennard, 2004, 2005 and Rebanks 2016). A deeply trusting, intuitive relationship between shepherd and dog is needed, and serves well to illustrate the expediency of trust. The well-known sheepdog handler and trainer Derek Scrimgeour puts it like this:

> Instead of adding additional pressure on your dog when he is already confused or worried you must be the one he relies on to make things easier for him to understand. If you can do this he will learn to trust you. In fact, he will begin to think he needs you. This is when a really special relationship begins to happen. (2002, 24)

I suggest that the rock art panel points to exactly such a special relationship.

Even though this rock carving is one of a kind, it sets the scene for this study of life in Bronze Age Rogaland. The motif testifies to the practice of both shepherding and pastoralism. The motifs of Bronze Age rock art emerge from the life-world and fields of experience of the society and *habitus* in which the artist(s) is embedded. In this book I will examine the archaeological and environmental evidence to investigate the stakeholders in that world and what parts they played. To gain a better understanding of the society that created this special rock carving, I will draw upon theories and approaches from Human-Animal Studies to explore how the animals themselves became active

where two dogs work together. He is an expert on handling sheep with the aid of dogs, and I have full confidence in his abilities to understand, interpret and evaluate the rock art motif from Valhaug.

agents and co-creators of their life-worlds. Further, I will investigate what kind of ontological status the three kinds of being on this rock carving held.

I approach the past from the perspective of multispecies archaeology, similar to Eduardo Kohn's "anthropology of life," which is "not just confined to the human but is concerned with the effects of our entanglements with other kinds of living selves" (Kohn 2007, 4). Further, I am inspired by Eben S. Kirksey and Stefan Helmreich and their multispecies ethnography, which centres on "how a multitude of organisms" livelihood shape and are shaped by political, economic, and cultural forces (Kirksey and Helmreich 2010, 545). My project, then, is a multispecies "archaeology of life" that investigates the interplay between living kinds of beings, both humans and other-than-humans and considers them as selves that partake in the emergence and shaping of joint lifeworlds.

Why this book

I have two main objectives with this book, the first one is driven by a need for theoretical exploration of human-farm animal interactions, and the second is my attempt to answer an archaeological riddle. First, I consider the lives of humans and animals in a particular historical setting as to allow for an investigation into the unfolding of culturally and historically situated ontological status of beings. Second, my research question springs from a very specific development in Early Bronze Age architecture in Northern Europe—the transition from two-aisled to three-aisled longhouses (Figure 1.2). Some archaeologists (Tesch 1992, 290; Rasmussen and Adamsen 1993a, 1993b, 138; Rasmussen 1999, 281; Lagerås and Regnell 1999; Årlin 1999; Armstrong Oma 2007, 2010, 2012, 2013b, 2016a, 2016c) believe this transition is connected to a shift to making room for animals indoors, whereas others dismiss such a claim based on lack of hard evidence (e.g. Carlie 2002; Petersson 2006). I will go into that discussion in depth in chapter 4. Suffice to say that when this happened—that animals became members of the household—I wonder why? I have never seen a line of argument that has solved this puzzle to my satisfaction. Often, functionalist explanations are given (see discussion in Zimmerman 1999b), but as explanations go they are simply inadequate, as will be explored in depth in chapter 4. I want to delve deeper than the mere functionalist, and examine the rationale underlying the kind of society that chose to alter their architectural layout to accomodate animals inside the house. Seeking an answer to this riddle has been a driving force that motivated me to offer this piece of work to the academic community.

To approach these matters, I have chosen to investigate Bronze Age settlement sites from Rogaland, southwestern Norway, because this area has been extensively examined both regarding its archaeology and, further, its environmental development and ensuing landscape changes in the Early Bronze Age. The case studies are remarkable in themselves, but even more so, they are interesting as an example of how to integrate Human–Animal Studies with archaeology. Also, this study serves as an example of how to study the lives of animals, and their relationships to humans, beyond studying their bones, since the case studies at hand offer varied archaeological data sets.

This book sets out a specific line of argument that concerns a few micro-level case studies, and with scant archaeological evidence. A valid question would be, do the case studies presented here really deserve so much attention and deliberation? I argue that there is a need to go into these issues in great depth. The specific way of understanding social interactions in the past that I strive for requires a thorough discussion of theory, and therefore I hope that this book will find its way to interested readers beyond those that work with Bronze Age Rogaland. By way of the case studies presented here, I aim to create a merging of Human–Animal Studies and archaeology, and from that particular perspective look at how archaeologists can understand and interpret the choices of people in specific places and at specific times in prehistory.

Figure 1.2 This reconstruction of a three-aisled longhouse shows the relationships between postholes, posts and roof frame, and how the post-built construction might have once looked when the house was standing. Illustration by Ragnar L. Børsheim / Arkikon.

An interesting avenue of further research would be to investigate if and how architectural choices in which space for animals was included in the building design has appeared elsewhere and at different times in prehistory. I hope that my study will inspire examinations of settlement contexts in other parts of the world with regard to how architectural choices were made to make space for both humans and animals. Here, I expand beyond Rogaland and look at some Early Bronze Age settlements from Jutland in Denmark to put the Rogaland Bronze Age society in perspective.

"The animal that therefore I am"

In a biological sense, the human species—*homo sapiens*—is one species amongst others in the mammalian kingdom—hence Derrida's admittance (2008). By its etymology, animal is related to *anima*—to breathe, to have spirit, and to be *animate* is to be alive and to move (Buller 2012, 142)—and have capacity for auto-motion. Acknowledging that humankind is inextricably linked to the rest of nature goes some way towards deconstructing the nature–culture divide that has dominated the humanities and led to a domination of studies on humans and their matters. Following a phenomenological approach, humans dwell in this world (see Ingold 2000b), and make up but one part of a much bigger picture. In dwelling, humans have made their home, together with the other elements and creatures that make up the world. Archaeology cannot, therefore, study past humans as though they alone inhabited the world, because their lives were seamlessly bound up with other creatures and their shared environment (Ingold 2000b). However, in the process of dwelling, people have taken different parts of their surroundings and creatures and used these to their benefit—as, indeed, do all living beings. This appropriation is what we term culture. I will therefore use this term, but bear in mind that it is not an opposition to nature—rather, culture signifies historically particular patterns of appropriation. Humans and non-human animals share this way of being in the world, simultaneously dwelling and appropriating. Often this has been done by some species side by side, and other times, in stark competition.

However, embedded in our language and culture there is a division between humankind and the broad animal category. Animals range from large predators, domestic animals to mice and rats, and sometimes also birds, fish and bugs. Humans relate to these various kinds of animals in numerous ways—from observing wildlife to hunting, killing, eating—and as companions, pests, food. Some animals are what Terry O'Connor (2013) terms commensal species (e.g. bacteria, parasites, insects and rats), others live in close relationships with humans.

> All our lives relate to all kinds of animals, from earthworms to beetles, from eagles to boa constrictors. For most people, however, the meaningful and enduring relationships with animal kin are those with companion or domesticated species, mostly mammals and birds—those in whom we may be able to recognize at least some emotional similarities.
> (Birke and Hockenhull 2012a, 2)

Derrida (2008, 34) also demonstrates that the phrase "animal" used to denote all kinds of animals is a profoundly unreasonable usage as it denies the proliferation of differences in animal lives, between for example lizard, sharks and lambs. Even though mice, rats, adders, fleas, lice, intestinal worms and wolves, moose, bears, wolverines, deer and foxes were probably prominently present in Early Bronze Age Northern Europe, I will focus on the domestic animals, those that have left physical traces in the archaeological record. Therefore, when I use the term animal I primarily mean domesticated animals that live together with humans, companion and/or farm animals. Categorical separations between kinds of domestic animals, such as farm animals (sheep) and pets (dog) were probably nonsensical to Bronze Age people. According to the sociological studies of Rhoda Wilkie, to modern-day stock persons, the separations between the categories farm animal, companion and pet are often washed out (Wilkie 2010). I suggest this was equally true for Bronze Age societies—like the humans themselves, the animals were probably perceived as workers—in accord with Camilla Årlin's (1999) suggestion that domestic animals were producers rather than products in the Bronze Age.

Animals in the past and today

Relations between humans and animals have, through the ages, been subject to changes in terms of engagement—both regarding practice and philosophy. Following these shifts in engagement, the lack of a firm boundary between humankind and animals has at times led to a deeply felt unease, since the boundaries between humans and animals have become contested. This is well documented for the Middle Ages in Europe, when animals were for example put on trial for having committed crimes (Figure 1.3; Fudge *et al.* 1999). In the words of Ericka Cudworth (2011, 21), "Our relations with non-human animals are historically situated—they have shifted in terms of the social forms they assume and the pre-dominant discourses which abound."

And yet, this irrefutable truth aside, how can archaeology contribute to Human–Animal Studies? Admittedly, it is difficult to fully grasp the relationships between presently living humans and animals, how then make valid statements about such relations in the past? The answer comes to us from the soil, and from the vestiges we discover of humans and animals that lived before us. Archaeology creates meaning by piecing together artifacts, whole and broken, settlements, monuments, graves and sacrificial offerings. Studying such objects and phenomena makes it possible to decode the practices that reveal belief systems, values and rationales that governed lives in the past. Often it is easier to see the wood for the trees from a distance, and as such easier to unpick the way animals were perceived and related to in societies that are not our own. We can, for example, gauge the physical and ideological scale of proximity and distance within the rationale of a particular society, and thus understand how human–animal relationships were allowed to unfold (see Armstrong Oma 2013b).

There is also a deep-seated connection between the lives of animals in the past, and those of today. This is true both in an empirical and imaginary sense. Empirical in the sense that there is a genetical continuity, and that choices regarding breeding and selection made in the past have formed the animals of today, both morphologically and temperamentally, and above all with regard to abilities. And imaginary, in the sense that the stories we imagine—and tell—about past animals, form the frameworks for how we think about current animals. But in fact, those tales we imagine are frequently founded in our current society's beliefs about animals.

One example would be the idea that the existence of farm animals is *a priori* to create nourishment for humans, and to disregard the wild ancestors of these

Figure 1.3 Illustration from Chambers Book of Days, "Trial of a sow and pigs at Lavegny." A sow and her piglets were allegedly tried for the murder of a child in 1457, the mother being found guilty and the piglets acquitted. By Unknown—The book of days: A miscellany of popular antiquities, Public Domain, https://commons.wikimedia.org/w/index.php?curid=19141170

animals as well as the process of taming, selection and domestication. Such narratives create a false sense of cultural and philosophical continuity from the past and into the present to perceptions of animals, their intelligence, personhood, agency and abilities (Armstrong Oma 2013c). Therefore, it is vital to the lives of animals today that we carefully consider those stories we tell. One example is the biologist L. David Mech who studied wolves in captivity in the 1970s, and from it drew conclusions about dog behaviour that has to a large extent shaped the dominance philosophy of dog training. He coined the phrase "alpha dog" that has made its way into everyday jargon. Unfortunately, he got the wrong end of the stick, and has later refuted his earlier research (Mech 1999). Making assumptions about wolves based on studies of captive wolves is akin to making assumptions about human societies based upon studies of refugee camps. In fact, he went so far as to put out a video on youtube in which he refutes his earlier conclusions.[3] Telling the "wrong" stories can have a huge impact upon the lives of animals. John Bradshaw demonstrates how this has been the case for dogs, many of whom have been treated with violence following advice from dog trainers who have usurped the (flawed) theory of hierarchy in wolves into leadership theories of so-called "alpha dogs" versus "alpha owners" (Bradshaw 2011).

3. http://www.youtube.com/watch?v=tNtFgdwTsbU

In this book I seek to establish ways of understanding past animal agency, also beyond how humans perceive animal agency. Animals and agency is explored in depth in chapter 3. Archaeology as a discipline is in a unique situation as it provides time depth to the question of how the agencies of humans and animals both have driven the course of history. This is for example relevant regarding environmental history, in which landscapes, environments and eco-systems have changed due to particular husbandry practices based upon social, cultural and economic choices. The methodology to investigate such questions is to perceive the archaeological record, which consists of material culture, as vestiges of lived, social practice, denoting how agents organized their lives and lived together (Barrett 2000), rather than as either a protocol into which objects should be organized, or material culture that could be read as a text (Patrik 2000). By exploring traces of practice we piece together fragments of past lives in the form of, for example material culture, landscape changes, architectural choices, burials and sacrifices, and so it is possible to decode the practices that reveal not only belief systems, values and rationales, but also the consequences of human–animal interactions that governed lives in the past. By using this perspective, and keeping a reflective frame of mind, I believe it is possible to go behind the cultural bias of the stories we tell ourselves about animals in our current society.

The role of animals in archaeology in the age of Humanism

The emergence of archaeology as an autonomous subject must be seen in light of the humanist *zeitgeist* of the nineteenth century (e.g. Thomas 2004). Archaeology is fundamentally an anthropocentric discipline, aimed at studying the human existence and human societies. Other elements that form part of human societies are seen as secondary objects of study, as a mere backdrop. Such elements are environmental factors, landscape, vegetation, climate, and also animals. The fundamentally anthropocentric attitude of archaeology is a trait it shares with anthropology, and the main focus of study in both of these is humans and their interactions. Even though other kinds of beings are not ruled out of inquiries *per se*, they are not discussed *as themselves*. Such other kinds are rarely considered beyond their role as a mirror to human society (Mullin 1999).

Zooarchaeology is a sub-field of archaeology that specializes in studying animals, mainly by way of their bones. This discipline grew out of the upsurge of environmental archaeology in the 1960s and 1970s (e.g. Albarella 2001; Grant 2002; Hamilakis 2001). As a discipline, it is informed by the natural sciences, that have tended to study animal behaviour, and have, historically, overlooked animals' relationships to people (Birke and Hockenhull 2012a, 1). The approach to animals most frequently seen in zooarchaeology, is in line with what Ken Shapiro (2008) describes as "Animal Studies." This approach opens up for the possibility that the humanities can pursue fields of studies beyond the human by seeing animals as a social construction, i.e. that their meaning, value etc is created by humans. Embedded in this way of studying animals is an attention to how humans will think about, and interact with, animals. But creating assumptions about what the animals themselves feel, think and perceive is seen as a merely speculative and thus pointless exercise. Any attempt at trying to enter the world of the animal is dismissed as anthropomorphism, i.e. the conveyance of human emotions, ideas and mind-sets into animal minds. One example of anthropomorphism is that animals that are pets will appear to be ashamed when they have done something the human guardian does not approve of. Within Animal Studies such deduction is seen to be nothing but a transferral of human emotion into the mind of the animal, and is totally unrelated to the inner world of the animal self. And since no deductions of this kind are accepted as valid, animals will forever remain objects of study only, and are not deemed as fit to enter inter-subjective relationships.

Animal Studies is coloured by its inheritance from behaviourism, and fundamental tenets in early twentieth century psychology and biology that claimed only behaviour could be studied and understood, whereas the mind was a black box, inaccessible to science. Archaeology, too, has developed within this scientific paradigm. Particularly processual archaeology was seduced by positivism and held that motives behind behaviour were beyond the grasp of archaeologists, but the behaviour itself could be observed, documented, measured and quantified. This led to some awkward circular argumentation, and Ian Hodder established that even though Lewis Binford did not wish to make statements regarding what went on inside the minds of people, he still made such tacit assumptions (Hodder 1992), and, I would add, these assumptions were culturally specific, based upon Binford's own modern, funcionalist rationale.

The perception of animals in zooarchaeology is therefore comparable to what Ken Shapiro identifies as Animal Studies. This approach rests firmly in the fundamental humanist tenet that places man at the top of the evolutionary ladder, as "top dog," as it were, and operates within the framework of a hierarchy of beings.

The idea that man is set to rule over the animals is deeply entrenched in the cultural history of Western societies. Historically, Aristoteles was the first authority to establish this as a "truth." The hierarchy of creation, with man on top, was established with inexorable certainty in Christian scriptures, and was preached every Sunday from pulpits throughout the Christian world. The church fathers, such as Augustine and Thomas Aquinas, all held opinions on this subject. Leslie Irvine (2004, 35–46) has a lengthy discussion about the role of animals in Western cultural history, and points to what a headache this question became to several of the Christian philosophers. She suggests that a guilty conscience is the root of the philosophically constructed ontological division between humans and animals. Instead of dealing with the discomfort of the desperate screams of dying animals, the priesthood dodged the problem by defining their way out of the disquiet by declaring animals to be soulless. Throughout the Middle Ages there was a long period of gestation of such ideas that culminated with René Descartes' famous *cogito* sentence. By hailing the powers of mind and thought and declaring the intellect to be the seat of the soul, Descartes elegantly redefined the bothersome question of the suffering of animals, and thus absolved humankind in the matter of killing and eating animals (confer also Franklin 1999, 2002). Keith Thomas (1984 [1983], 34) argues that the most powerful argument of the Cartesian position is that it was the best possible rationalization for the way people were actually treating animals.

The philosophy of Descartes was instrumental to the development of the era of enlightenment and humanism, in which intellectual progress and science were seen as driving forces that would improve the plight of humankind. These ideas have been formative to the ethos of Western science, and are deeply seated in the "hard sciences." In addition to striving for intellectual progress, enlightenment ideals are the supremacy of man and a hierarchy of creation. Althogether these principles became fundamental to the enlightenment project, and thus deeply structuring to the evolution of modern science, and also to the development of archaeology as a discipline.

Currently, a deep-felt unease is emerging about the philosophical and ontological perspective of animals that is embedded in this part of science. Many researchers from different disciplines feel that this perception of animals is flawed, and are working in the interdisciplinary field of Human–Animal Studies, that challenges the hierarchical model with humans on top.

Human–Animal Studies and archaeology

Human–Animal Studies is a newly established interdisciplinary research field that has had an immense growth the last few years (Shapiro and DeMello 2010). Several journals are devoted to the field, such as *Society and Animals*, *Anthrozoos* and *Humanimalia*, and Brill publishes a series called *HAS* (short for Human–Animal Studies). Large international conferences are held, for example under the umbrella of the loosely knit organization Minding Animals.

The objective of Human–Animal Studies is to examine the complex and multidimensional relationships between humans and other animals—and it "embodies a sustained interest in understanding and analysing how we humans relate to and make sense of other species" (Birke and Hockenhull 2012a, 3). Human–Animal Studies comprises work in areas of study such as psychology, sociology, anthropology, political science, history, literary criticism and other disciplines in the humanities.[4] A plethora of research questions emerge from this platform, examples range from animal-assisted therapy, wildlife conservation, to questions of animal advocacy and rights. An outspoken position of animal advocacy is frequently found, DeMello and Shapiro (2010) point out that the combination of advocacy with scholarship is currently perceived as less threatening, as the thrust of contemporary philosophy of science challenges the traditional view that it is possible for a scholar not to have a position on the material under investigation. The thing is, as Birke and Hockenhull (2012a, 4) point out, understanding multiple forms of human-animal relationships "requires thinking laterally, to interrogate widely diverse ways of observing and theorizing, and to be able to draw on diverse disciplinary approaches." This broad platform includes, for many, a personal investment with the issue at stake.

For a large part, the archaeological literature on animals is, as previously stated, within the bounds of Animal Studies. However, there are exceptions—namely archaeologists that work within Human–Animal Studies (see for example *Society and Animals* volume 21, issue 2 from 2013, which is a special issue on archaeology). Human–Animal Studies allows the animals to be objects of study, but they are also sanctioned as subjects. Academic works within this field own up to the fact that for animals to be studied as social constructions—and to understand the meanings of those constructions—the question of who and what the animal in itself is must be addressed first, such as their social and cognitive abilities and behaviour (Shapiro 2008; Shapiro and DeMello 2010). Meaning emerges through experience, for human and non-human animals both; also in the way humans understand and interpret the inner world of animals. One example is how the relationship between humans and horses is shaped. The material culture

4. See https://www.animalsandsociety.org/human-animal-studies/

that the relationship is performed through creates a joint framework for ways of being together. Furthermore, the material culture and its embedded framework is conceptualized based on preconditions within the horses (Birke *et al.* 2004). This necessitates certain knowledge of, and feel for, the horse itself to reach an understanding of the horses' function, participation and/or symbolic role within a community (Argent 2010, 2012; Armstrong Oma 2011, 2013a). For example, Gala Argent's poignant study of the Iron Age Pazyryk people and their burials suggests that horses were dressed up according to their biography and particular life events (Argent 2010).

Recent research on biochemistry has demonstrated that it is indeed possible for humans to have an understanding of what it feels like to be an individual of another species, by way of the biochemical phenomena of the hormone oxytocin (Handlin 2010; Olmert 2009) and brain cells termed mirror neurons (Iacoboni 2008; Rizzolatti *et al.* 2009). These phenomena, and their implications for the understanding of human–animal relationships, are discussed in chapter 2.

Animals in archaeology—potentials and pitfalls

Archaeology holds the potential to be a very important discipline for the development of Human–Animal Studies since archaeology can add a unique diachronic perspective, exploring the co-evolution of species, and adding time depth to issues that concern Human–Animal Studies (Armstrong Oma and Birke 2013). In the deepest sense, archaeology can, by way of studying human origins, ask fundamental questions about what it is to be a human animal, and thus de-centre the human species, pushing us off our self-appointed throne of species supremacy (see Gittins 2013).

Further, archaeology can give depth to perceptions of animals not only across cultures and space, but also through time. Examples range from hunter–prey relations, to living with animals, both as farm animals and companion animals. Promising topics in this regard are the construction of boundaries between humans and animals, for example through art, mortuary practices and domestic arrangements. Such investigations pave the way for more deeply-set questions regarding ontology (see Brittain and Overton 2013).

Another area in which archaeology can provide an incomparable contribution is the cultural context of domestication. Understanding the processes of domestication of animals goes beyond simply taming and ensuing genetic changes in animals, but addresses new ways of living, cross-species, both politically and socially (see Argent 2013; Mlekuz 2013). Also, archaeology can nuance the understanding of the history of domestication as a gradual process (see chapter 6 for further discussion).

However, those outside of the discipline question archaeology's contribution to such questions. For example, anthropologist Natasha Fijn (2011, 22) questions the theoretical tenets upon which archaeology is based:

> Animal domestication is often investigated in the realm of archaeology, especially zooarchaeology, in an attempt to establish the origin of domestication in animals. This line of questioning could be flawed from the outset in that the aim is to seek a point where humanity transcended nature and was set along the path towards culture and civilisation. Anthropological thinking is moving towards the idea that humanity has never been separated from nature...

Following Ingold (2000b), she argues that the same is true for animals, stating that we cannot pinpoint an exact point when a species goes from a state of being wild—and part of nature—to one of being domesticated—and a part of culture (see also Shanklin 1985; Zeder *et al.* 2006). That point in time simply does not exist, but is rather an ongoing mutualistic relationship where both partners become interdependent.

Such critique emerging from outside of archaeology should be taken seriously. What is further problematic is that this kind of critique is often justified. For example, in her latest book, *Social Zooarchaeology*, Nerissa Russell reduces the complex question of domestication to one of ownership: "I see the key quality of domestication as social: Domestic animals had owners" (Russell 2012, 248). In view of Fijn's line of argument above, Russell's statement is simplistic if for no other reason than it is impossible to pinpoint the moment in time in which humans came to regard themselves as owners of animals.

Studies of animals in archaeology—bones and calories

A quick glance at the studies of animals within archaeology itself can shed light on the reticence to deal with Human-Animal Studies found within the discipline. To understand the underpinnings of this resistance, the development of zooarchaeology is a useful starting point. Zooarchaeology is a recently developed discipline, traditionally placed within environmental archaeology, a highly specialized branch of archaeology that has tended to isolate itself from mainstream archaeology (Albarella 2001; Thomas 1996, 5; Overton and Hamilakis 2013, 112). The aim of environmental archaeology was to create methods that allowed an understanding of past environments, that within this framework of research comprised a backdrop to human societies. Ecological factors such as environment, seasonality and behaviour were at the core of the study of animal remains (Jones 2002, 131). Processual archaeology, the scientific outlook which emerged in the 1960s, thus appropriated animal bones as a valid area of study (Binford 1978; Thomas 1996), and set the standard for what animal bones could tell, in the words of Annie Grant:

> [m]uch of the research that begins with the identification and analysis of animal bone remains from archaeological sites has concerned itself with the reconstruction of husbandry practices and patterns of consumptions at subsistence level. This is perhaps unsurprising as the development of archaeozoology was very closely linked with the "economic archaeology" of Grahame Clark and Eric Higgs. (Grant 2002, 17)

I support Grant's conclusion that archaeologists with direct experience of animal husbandry as it was practiced in the functionalist-minded 1950s and 1960s enforced the harsh realities and adaptive pressures of stock management as narratives that they held to be valid for past societies.

These kinds of zooarchaeological studies seem to be more deeply embedded in the frame of mind of today's society rather than that of the societies of the past. Their approach is heavily influenced by the cost-benefit maximization of modern farming, in which an enormous pressure to create cost-efficient solutions lead farmers to see animals as products. This is the consequence of two modernist trends. First, the removal from the reality of the human-animal relationship in modern meat production, in Franklin's words:

> our relationships with farm animals have become increasingly separate as people of the twentieth century have become spatially detached from the animals they consume and emotionally reluctant to recognize the embodied nature of consumed foods. (1999, 126)

Second, farm animals have become almost exclusively a source of protein, whereas in most pre-industrial societies animals are primarily significant as a source of labour (Franklin 1999, 126–127), i.e. as producers rather than products (see Årlin 1999). These factors explain why traditional zooarchaeology fails to recognize animals as having a life and a place in society before they were eaten (pathology that demonstrates traction is the exception). This might account for why animals have been dealt with in a rather simplistic way, in which the human–animal relationship has been considered to be a one-way influence, where humans exploit animals (see also Ingold 2000a). Wider implications in terms of ways of living with animals, and implications of different regimes on the social and/or economic practice of the society at large, are never considered.

Critique of zooarchaeology...

This approach has been heavily criticized from within, as several zooarchaeologists (e.g. Albarella 2001; Grant 2002; Hamilakis 2001; O'Connor 2001) have heavy-handedly hammered away at both methods and aims of their own discipline: "these reconstructions are like publicity stills from a movie: they show the sets and the props, but reveal nothing of the plot, still less the script" (O'Connor 2001, 20). Zooarchaeology is thus not relevant to the broader questions addressed by archaeology: "environmental archaeology is trapped between the straitjackets of methodology—science—and a thematic investigation—the environment" (Albarella 2001, 7). Or, put another way, "[e]nvironmental archaeologists, therefore, find themselves between the rock of empiricism and scientism, and the hard place of social constructivism and western dichotomous thinking" (Hamilakis 2001, 30).

Going "beyond subsistence" (Grant 2002, 17) has not traditionally been seen as a possibility. For example, site-catchment analysis (e.g. Higgs 1972) assumes that all available resources that are edible will be eaten. It fails to take into account that people are cultural beings with culturally informed, historically situated, perceptions of what constitutes food. However, even the most adamant of functionalists admits that cultural factors must also be taken into account regarding food consumption (Bökönyi 1975).

But is it all about food? Bradley states (2005, 8): "[a]rchaeology is impoverished unless it sheds its fixation of food production." Similarly, Whittle (2003, 82) argues: "[i]n place of notions of value, belief, categorization, symbolism, time and gender the principal concerns in the archaeological field have been practical, economic and evolutionary."

However, in spite of the internal critique, the zooarchaeologists themselves have not had much success when it comes to proposing alternative frameworks. A reinvestigation of methods (e.g. Marciniak 2005) and aims (e.g. Albarella and Serjeantson 2002; Grant 2002; Halstead and Isaakidou 2004; Halstead and Barrett 2004; Hamilakis 1999; Hamilakis and Konsolaki 2004), have led to a change of focus from animals/meat within functionalist domestic economy to animals/meat into the social context of eating, most notably feasting (for example Dietler and Hayden 2001; Halstead and Isaakidou 2004;

Hayden 1996; Pappa *et al.* 2004). Some complexity has been gained, but rather than in-depth studies of animal meat and its calorific value, animal meat is now placed in the context of social status. Apparently, it is still mostly about food. Also within the social complexity paradigm, then, animals are not considered as themselves. For example, in Russell's (2012) overview, she concentrates on topics such as hunting, rituals, symbols, animals as pets, feasting, with a very short section on human–animal relationships at the end. In all of these, her focus is on how animals have been used by humans. Overton and Hamilakis' (2013) discussion article on social zooarchaeology leaves me more hopeful in terms of a future for the discipline. In it, swans that were hunted in the Mesolithic are considered as themselves—as sentient beings, agents with a life-world of their own, and at the same time elements of economizing and exploitation are considered. This work thus balances the consideration of what actions the bones represent with the beings from whom the bones stem.

...and the dilemma

Why is it so hard for zooarchaeologists to engage with their object of study beyond culling patterns and butchery marks? For the archaeologist, the difficulty arises when we meet our study object—the animal bones. Although we intellectually know that the bones stem from once-living animals, the leap of imagination required to reconstruct the life of that animal is frequently lacking. The very nature of the working conditions of the zooarchaeologist goes some way to explain this lack. The time pressure involved with the post-excavation work often leads to a scenario where deadlines preclude analytical work that goes beyond zooarchaeological analyses. Further, the bone material given to the zooarchaeologist can be very fragmented, leading to an analytical mindset in which the prevailing bias is to scientifically analyze the bones according to the environmental information they can give. For example, often only small pieces of unidentifiable splinters remain. In other cases, the bones are less fragmented, but might have clear butchery marks. The carcass itself may have been fragmented, so that just one bone from each individual animal is present. In a run-of-the-mill excavation, maybe 500 bone fragments are retrieved, 200 are identifiable to species and skeletal element, and 30 can be identified to sex and age. In such a scenario, each bone fragment might represent an individual animal—or several individuals might be found within the assemblage of bones. The uncertainty implicit in the fragmentation is not conducive to leaps of imagination regarding the lives of those animals. Andrew Shapland says simply (2013, 204) "it is difficult to make the animals of the past come alive"—despite his object of study, a figure depicting a bull and a so-called bull-leaper. How much more difficult is it not to make bone fragments come alive?

In my view, the critique from within has so far done little to radically alter zooarchaeology, and a new framework is in order. To fundamentally develop an archaeology that considers animals, the animals must be considered as themselves, rather than as mere flesh and bones. To do this requires courage, imagination, and an insight into Human–Animal Studies. Notably, those that try to carve out new frameworks for animals in archaeology come from outside of the zooarchaeological tradition (Argent 2010, 2012; Årlin 1999; Armstrong Oma 2007, 2010; Mlekuz 2013).

Animals as symbols

Those that have ventured beyond the bones have tended to focus upon the symbolic aspects of animals, influenced by the post-processual archaeology of the 1990s, whose clarion call was a search for meaning in the archaeological objects themselves. Objects should be read as though they were texts—regarding animals, this implied a symbolic reading of them (e.g. Hamilakis 2003; Hamilakis and Konsolaki 2004; Hedeager 1999; Hill 1996; Kristoffersen 1995). The study of animal symbolism discloses how their meaning is constructed in culturally specific ways, for example the trusty dog or the peaceful dove. Sometimes these culturally specific meanings tell us more about the society in which they arise than of the animals themselves. For example, Cudworth (2011, 5) has recently pointed out that dogs are trusty companions in the Western world, and meat in China. And cows are meat in the Western world, but holy in India. In part, the quest for symbolism in archaeology has taken its cue from anthropology, inspired by the much-quoted phrase by Claude Lévi-Strauss, "animals are good to think with" (1964 [1962]). This widely accepted premise has led to a one-sided focus on the representations of emblematic animals in both anthropology (Milton 2005) and archaeology (Armstrong Oma 2007, 23).

However, I argue that this approach did not have much more of a success in bringing the animals back to life than the zooarchaeologists—rather than considering animals as themselves, they became subject to a reading of their culturally specific symbolic meaning, in line with what Shapiro (2008) would call the social construction of Animal Studies. Whereas reading for meanings can sometimes give fresh insights into many aspects of society, it should not be mistaken for a study of animals as themselves. All of the "thinking with animals" has led to reducing animals to a mere mirror of human society (Mullin 1999), rather than studying animals as part of those societies. Archaeology is not the hand maiden to anthropology, and the particular ramifications of archaeology adds time-span to anthropology's cross-cultural width. Working both diachronically as well as synchronically gives some other ramifications that set archaeology apart, that provides the scope to study change, processes and developments. Nor is archaeology the hand maiden to history, since archaeologists study the material remains of the past—the actual physical lives—rather than the stories those people told about themselves.

Were they all human?

Archaeology is in the singular position that the objects that we study are the real thing, the lived lives. Animals partook, one way or another, in those lives. The soundscape of the past included sounds made by both humans, animals and their environment, likewise the smellscape was formed by all beings that inhabited the lifeworld. The point I wish to make is in some ways similar to the feminist slogan "were they all men?" Obviously, animals of all kinds, shapes, sizes and colours were ubiquitous in the past, but their power to impact societies as themselves rather than as mere food for thought —"good to think with"—or sustenance for the body—"good to eat"—has so far been grossly underestimated (Mullin 1999). This way of thinking about, rather than staying with (in the words of Donna Haraway, "staying with the trouble"[5]) animals has led to a

5. https://www.youtube.com/watch?v=GrYA7sMQaBQ

particular view of animals: I cannot find words to express this better than John Knight has: what has happened is

> collapsing the two separate dimensions of the human relationship to domestic animals —outcome and process. The preoccupation with the outcome of the relationship (e.g. slaughter for meat) is apt to conceal the protracted relationship of nurturance and care that precedes it. (Knight 2005, 5)

This book seeks to investigate in depth how a Human–Animal Studies approach can inform archaeology and transform the understanding of the past, not only in terms of the roles of animals and their relations to humans, but to more deeply inform scholars of the workings of society. To fully understand the workings of society, anthropocentrism must be challenged. A core idea in my work is to investigate how such an approach works within the archaeological discourse—and explore ways in which it can work.

The time has come to trace the lives of those animals, their biographies. Human–Animal studies provide the theoretical framework for doing so. This work is a contribution to Giorgio Agamben's call to "smash the anthropological machine" (Agamben 2004). This book is dedicated to exploring avenues of research into what John Knight (2005) calls "the protracted relationship of nurturance and care" and thus staying with the animals in the Bronze Age world.

— 2 —

Understanding Animals: Perception, Sentience and Anthropomorphism

Animal sentience can inform studies on the past

How can animals be understood as themselves—and how have they been understood through history? This chapter draws upon stories told about animals through time that disclose how perceptions of animals have been different in the past compared to today. To investigate the understanding of animals in past societies I present new advances in science that provide novel avenues for fathoming animals in different ways. The discussions in this and the following chapter might not initially seem relevant for archaeology, but it is important to establish *what animals are* in order to investigate how they were understood in the past, and how archaeology can understand the role of animals—and thus the formation of past societies—in their fullest sense. I make a case for the sentience of animals, based upon new directions within the interdisciplinary fields of Human-Animal Studies (HAS) and Animal Studies (AS), as well as biology. This has implications for the theoretical arguments that unfold in this book.

In the following chapter I argue that animals, by their true nature as sentient beings, are agents and can be understood as such, not just by people today, but also by people in the past. This has implications for how past societies are understood, such as how societies took form, and according to which structuring principles, and by the power of which agents. By this I mean that animals could also, by their ability to act, negotiate and thereby influence the course of history—even though this might not have been their intention. Second, seeing animals as sentient beings has implications for how humans act towards, and co-create their lives with, animals. Therefore, animal sentience has implications for interpretations of past societies.

To make a case for this reasoning—ways of perceiving animals, their agency, minds and emotions—I will outline the discussion of anthropocentrism in this chapter to create a framework for how, and to what extent, humans can understand animals. In the following chapter I argue that the sentience of animals means that animals are agents; they have the ability to reciprocate, and not just by literally sinking their teeth into the experimenter, but also in social ways, for example by cooperating, resisting, or striking out a new path.

Two separate arguments, that have a bearing upon each other, will be presented in this chapter. The first is informed by Human–Animal Studies and engages directly with the question of animal sentience and therefore the animals as themselves. However, whether or not human animals perceive themselves as part of the mammalian kingdom and grant other animals sentience, is another question. Therefore, the second argument is informed by anthropology, sociology and the so-called ontological turn and sets out to explore how sentient animals are understood in different human societies. In this chapter I will go back and forth between these two arguments, since they are tightly interwoven considering that the human perception of animal sentience, and its consequences, is key to the interpretation of the archaeological data. This is not a treatise on biology or neurology; therefore I trust the findings of those disciplines and will mostly explore the second tenet—how can humans understand sentient animals and what implications does this have for the way that humans and animals co-create their life-world.

I do not, therefore, imply that animals were always and everywhere perceived as sentient by humans, and that people in the past always had awareness of animal sentience. But given that those people, especially in farming societies, lived lives that were inextricably and intimately woven together with lives of animals, there is a great potential for the recognition of animal sentience. In fact, I will argue that the material records in my case studies demonstrate that humans have indeed recognized and acknowledged the animals with whom they shared their lives as sentient beings.

Animal sentience

In current research, there is an increasing awareness of the sentience of animals that has led to advances in research on "minding animals" (Bekoff 2002), a word play in three parts: it takes into account animal minds, gives attention to the animal question, and implies caring for animals. This recent development in animal studies has led to the need for publicly stating that there is now a consensus amongst scientists that animals are sentient beings with consciousness. In the Francis Crick Memorial conference in Cambridge in July 2012, the finest brains on the brain, e.g. neuroscientists, biologists, etiologists and psychologists issued the so-called Cambridge Declaration on Consciousness.[1] In it they state:

> The absence of a neocortex does not appear to preclude an organism from experiencing affective states. Convergent evidence indicates that non-human animals have the neuroanatomical, neurochemical, and neurophysiological substrates of conscious states along with the capacity to exhibit intentional behaviors. Consequently, the weight of evidence indicates that humans are not unique in possessing the neurological substrates that generate consciousness. Non-human animals, including all mammals and birds, and many other creatures, including octopuses, also possess these neurological substrates.

The biologist Marc Bekoff's poignant comment was that the statement left him somewhat baffled as this is, in his words, "something we already knew" (Bekoff 2012b), and it can seem like stating the obvious to those that work and live with animals. However, it is positive that scientists are increasingly recognizing the sentience of animals. A steady in-flow of new studies assert animal sentience in a range of species, from earth worms to octopuses.

1. http://fcmconference.org/img/CambridgeDeclarationOnConsciousness.pdf

At an animal welfare level, research on animal consciousness can have an impact. For example, within the legislation of the European Union, the Lisbon Treaty of 2009 states that animals are sentient beings, and welfare legislations take this into account.[2] It is to be hoped that acknowledging the sentience of animals will also trickle down in archaeology, and I hope that this work will contribute to that.

Since the objective of archaeology is the study of the past, it goes without saying that archaeology does not engage with these questions directly. The result is that archaeology is often stuck in an old-school behaviourist world-view in which animals are automata governed by instincts, a lingering inheritance from the advent of archaeology as a fundamentally humanist project (as discussed in chapter 1). In this parsimonious understanding of the nature of animals, animals are not capable of consciously acting within their environments; rather, they re-act, in a stimulus-response fashion. This reductionist approach fails to acknowledge that many animals can be seen to move beyond passively or instinctively reacting, to responding and initiating. Obviously, they can choose to do so in ways that are either beneficial or dangerous to humans. They can choose to bite back, or they can choose to bond and liaise with humans. The sheer mass of animal studies that every day add more weight to making a case for animal consciousness, and to deepening our understanding of the natures of our conspecifics, will in time drive the final nails in the coffin of what Anna Tsing and Donna Haraway call the fantasy of human exceptionalism, which has blinded the academic community until recently (Haraway 2008; Tsing 2012).

My approach is multidimensional, comprehensive and opens up for different ways of perceiving and approaching animals. A compartmentalizing approach, such as seeing animals as either calories or symbols, narrows the perspective so that too much is "lost in translation" because such research cannot do justice to the complexity of the original situation (Armstrong Oma 2008a, 2012). Often, this point of view is mistaken for a romanticizing, sentimentalizing kindness. But, in the words of Donna Haraway, "We are face-to-face, in the company of significant others, companion species to one another. That is not romantic or idealist, but mundane and consequential in the little things that make lives" (2008, 93). Romanticizing animals is not my intention—rather, I want to open up a broader field, where concerns such as relationships, affinity, love, welfare, care, trust—and even kindness, are brought into the debate, alongside more hard-nosed concerns, like economics, killing and subsistence. Anna Tsing (2012) puts it thus: "Domination, domestication, and love are deeply entangled. *Home* is where dependencies within and among species reach their most stifling." Most of all, I argue that this field is about *practice*, the way real people and real animals as emergent entities have and lived their lives together and co-created their lifeworld (Armstrong Oma 2007, 2010).

Form versus content

It is necessary to take a closer look at the reasons why archaeology, and in particular zooarchaeology, have disregarded sentience in species other than humans. Traditionally, archaeology as a discipline has been obsessed with form—for example the shape of an object, its decoration, use of materials, in short—how something looks, be it a

2. http://ec.europa.eu/food/animal/welfare/policy/index_en.htm

pot, a brooch, or a house plan. This is no wonder since our evidence consists of material culture, which is often in a fragmented state. This gave birth to typology, and has been an important tool in ordering the archaeological record, in an attempt to create a place for that specific pot, brooch, etc, in a linear time-line (Trigger 2006). Content has been discussed considerably less, although post-processual and gender archaeology has significantly improved the theoretical tool kit of archaeologists. Often, it is felt that we cannot comment upon the content, because we haven't got access to the minds that thought into life the content or meaning of an object. However, in real life there is no such dualism between form and content. The meaning is the form and the form is the meaning—or, as archaeologist Siv Kristoffersen (1995) has suggested, the meaning of the object penetrates the surface of the object (inspired by Lévi-Strauss 1963). Considering animals, this bears a particular importance. Zooarchaeologists study bones purely on the basis of form. Form of bones gives information about species, body part, possibly age of death and sex (Reitz and Wing 1999). However, the bones also represent what could be called their *content*, namely the animals that they once were. This is easier to grasp regarding human bones, we recognize the sanctity of the human life that once incorporated the bone. I suggest that we should grant animal bones the same consideration—they are both meaning and form. Archaeologists have reiterated a false dualism, and regarding animals, this has had a negative effect. Giving privilege to form over content has meant that the lives of the animals that are represented by the bones have not been considered, except to tease out the cull ratio and the sex-age-ratio, hard facts that are used to figure out an economic strategy.

Regarding the lives of animals, we need not penetrate the minds of the people of the past—but of those lives of the other species whose bones we find. However, as with humans, this feels like an insurmountable task. Not only penetrating those eons of thought, but also the species barrier! However, some newer anthropological research has considered the lives of humans and animals, their intersubjective relationships and questions related to ontological status beyond the species barrier. It is to this research that I turn, in order to unravel potential relationships between past people and the animals that shared their lives.

Insights from anthropology—a multiplicity of worldviews

Archaeology often turns to anthropology when faced with questions of an inter-subjective nature. However, when it comes to animals, anthropology has not fared much better than archaeology, and has—as mentioned previously—tended to study animals as mirrors of human society (Mullin 1999). There are, fortunately, notable exceptions (see for example Kohn 2013; Morris 1998, 2000; Strathern 1988; Thompson 2011; Viveiros de Castro 1998; Willerslev 2007), and I will present some of the anthropological literature on personhood, since animals and also objects are sometimes given personhood status. This creates ramifications that have some implications for how animals are perceived.

A word on the accounts of indigenous peoples before we proceed. Frequently, ethnographic accounts give the environment, and different elements of it, agency. One such account is Hallowell's study of the Ojibwas (see 2010), made famous by Ingold (2000b). Hallowell recounts how animals and objects (even certain stones, as Ingold is fond of reiterating) were perceived to have agency. However, this goes against the grain of the western metaphysical principles—so what to make of such tales? To the Western scholar, it is an impossible idea that a stone can actually have physical agency in the real world. The Western scholar solves this insurmountable gap between her worldview and that of the indigenous people by suggesting that the indigenous people are really speaking metaphorically, "as if" (Willerslev 2007, 2–3). Thereby, the western metaphysical view is given primacy over the stories that indigenous people tell about themselves. The subtext for the Western scholar is that indigenous people have misunderstood reality, and reality needs to be translated by western anthropologists, into their own supreme worldview.

Conversely, the anthropologist Rane Willerslev seeks to understand the stories of the people he studies, the Yukaghirs, at face value, and gives primacy to the interpretations and understandings given by the storytellers. Rather than giving priority to the Cartesian dualism that Western science is entrenched in, he wishes to reverse the situation and give supremacy to the Yukaghirs, trusting that they are best equipped to narrate and translate their worldview. By successfully drawing upon Heidegger's *dasein*—"being-in-the-world" (1962), Willerslev succeeds with understanding and explaining the worldview of the Yukaghirs, such as personhood attributed to animals. The world is given meaning according to direct perceptions of it, that take place simultaneously with engagement. Here, sentience itself merges with the perception of it.

Thus, Willerslev (2007) suggests an approach that is fundamentally phenomenological, based upon Heidegger's concept of "being-in-the-world," emphasizing both "in" and the hyphens that bind the words together, forming a relational mode of being. This differs fundamentally from the Cartesian separation of the body from the mind, in that the mind is fully embodied and does not exist without a body. Willerslev argues that this approach untangles the precepts of traditional anthropology—and the mess they create, in which the accounts of indigenous people are rendered as misunderstood by the peoples themselves.

Like the Ojibwas, the Yukaghirs are hunters. They perceive animals, and prey, as possessing personhood. The personhood of particular beings is born out of a direct engagement and relationship with them. Thus, not all beings are given personhood. Rather, "personhood is constituted in and through the relationships into which they enter" (Willerslev 2007, 21), and is as such a potentiality of their being in the world. "An animal therefore can be just that, or it can be a subject-person with a mind of its own" (Willerslev 2007, 21).

Such insights from anthropology have to some extent, and successfully so, been incorporated into archaeological studies. Jones (2009) points out that a Cartesian separation is unlikely to have existed in the worldviews of prehistoric people. Most often, this perspective has been integrated in studies of hunter-gatherers (Conneller 2004; Gittins 2013; Hill 2013; McNiven 2010; Overton and Hamilakis 2013). There is a budding consensus amongst Stone Age archaeologists that the worldview of past hunter-gatherers must have been "pre-Cartesian." Not so with studies of farmers. Here, a functionalist ethos prevails, that assumes that prehistoric farmers were largely harbouring the Cartesian separation as their mental world-map (Orton 2010, 189). However, what would the consequences be of assuming that farmers, like hunters, perceive other beings as possessing personhood? Moreover, unlike hunters who engage fleetingly and tangentially with the animals they attribute personhood, the accumulation of time, engagement and practice shared between farmers and domestic animals would even more so allow for relationships to develop and for the animals to move beyond the "animal category" and become that sheep who is so stubborn, or the cow who is such a great mum, or that dog with the snappy sense of humour—thus seeing each animal as a, in Willerslev's words: "subject-person with a mind of its own" (2007, 21).

It would appear that also scientists (for example those contributing to the Cambridge Declaration on Consciousness) support the notion of animals as beings that could be imbued with personhood, depending upon the definition. An important question to address is, nonetheless, how those that engage with animals come to understand those animals, what is their understanding based on, and are they anywhere near a true understanding of what is actually going on in the minds of animals? How do humans see and evaluate the personhood of other species?

Anthropomorphism, egomorphism and sense of self

A poignant discussion amongst biologists and ethologists is whether animals have a sense of self, and how humans can recognize the animal self. Ironically, most humans who have had dealings with animals outside of the lab (especially with companion animals) will testify that animals have a sense of self. For example, I could personally relate a number of stories about my companion animals, such as when my border collie Bessie looks at herself in the mirror and doesn't bark at her reflection, thus recognizing the mirror image as herself.[3] However, such stories are to a great extent dismissed as "anecdotal evidence" by biologists (discussed in Bekoff 2007, 121). The objection from the hard sciences is that such situations cannot be recreated and quantified in the lab, and thus animals' sense of self cannot be proven, scientifically.

An understanding of animals' sense of self was traditionally thought to be mere human mind games, referred to as anthropomorhism, i.e. the attribution of human states of mind onto animal minds. Even though there is now a large body of data that supports the existence of a sense of self/sentience in most animals, the difficulty remains as to how to understand and interpret animals. Anthropomorphism is the tool that scholars apply to talk about animal minds and sentience. An in-depth discussion of ways of perceiving animals is important to the case studies presented, since perceptions of animals in past societies matters to the arguments that unfold.

Social scientists disagree regarding both the applicability of the anthropomorphic point of view and its value. Anthropomorphism is a common analytical tool when perceiving animals, but is demonstrated by Milton (2005) to be fundamentally flawed. In the social sciences, anthropomorphism tends to be used in one of the three following ways to describe preconceptions of animals in modern societies (Milton 2005, 256–259; see also Bekoff 2007, 156):

- Animals are represented in myths, fables, stories or cartoons as animal-formed representations of humans; in which animals are zoomorphic humans that think human thoughts and live and dress like humans, such as Mickey Mouse (e.g. Minahen 1997).

3. The so-called mirror test is used to demonstrate that primates have a sense of self.

- People in the modern world treat animals as though they are humans, such as pets that are dressed in clothes, wear jewellery and may receive Christmas presents.
- Animals are thought of as "persons," and are attributed human characteristics and mental faculties, such as having motivations, intentions, and emotions which humans recognize from their own experience as inseparable from being human.

However, Regan (1988, 6) points out that although anthropomorphism is understood to mean the attribution of human characteristics to things that are not human, the logical meaning must rather be to attribute characteristics that *only* belong to humans, to things not human.

Anthropomorphism in the social sciences originates, according to Milton, from a reluctance to make assumptions about the "inner" states of animals. But in doing so, another assumption replaces it, namely that animals have no "inner" state at all. Thus, a lack of internal coherence is exposed: in order to describe the way animals are understood as anthropomorphic, the opposite of what anthropomorphism itself implies should be assumed, namely that animals are not capable of the inner states supposedly attributed to them, as these are human characteristics (Milton 2005, 259). Milton suggests that personal experience, rather than generic human-ness, should be the basis for understanding others, "and that understanding is achieved by *perceiving* characteristics *in* things rather than, as anthropomorphism implies, *attributing* characteristics *to* things" (Milton 2005, 260, original emphasis). Milton suggests that an alternative tool for perceiving animals is *egomorphism*, which is the understanding of others (both humans and animals) according to whether or not they are "like me," rather than "human-like." Thus, there is no difference between the perceptions of fellow human beings as persons, and the perception of animals as persons. When we see outward signs of emotions, we immediately infer an inner world of emotions, whether it be a human or an antelope (Milton 2005, 260–265). The distinctions suggested above highlight the need for different ways of understanding human–animal interactions, intimacies and relationships. Egomorphism as an analytical tool suggests that perceptions of both self and other are formed through relationships.

Alternatively, the innovative and radical ethologist Marc Bekoff holds that anthropomorphism is a useful tool, and coined the phrase "biocentric anthropomorphism" that denotes "taking the animals' point of view by using the information that we have about the world of members of that species" (Bekoff 2007, 40). Bekoff refers, sarcastically, to anthropomorphism and anecdote, together with analogy, as the "nasty A-words" (Bekoff 2002, 103; 2007, 113). The second nasty A-word, anecdote, is dismissed because anecdotes are not "hard data" and cannot be reproduced in the lab. According to Bekoff, many scientists see anecdotes as "tainted with personal involvement and biases" (Bekoff 2007, 121). But quite to the contrary, Bekoff argues that anecdotes are vital to ethology, and—I would argue—to the social sciences in general, because systematic analyses of anecdotes can lead to data that are reproducible by mimicing the anecdotal situation. I therefore agree with Bekoff's claim that "the plural of anecdote is data" (Bekoff 2007, 121). Regarding archaeology, one could argue that all evidence is in a sense anecdotal, since no hypotheses can be reproduced in a lab—it depends on what we happen upon during excavations.

Figure 2.1 The intense gaze of a young fox. Photograph by Jonn Leffmann, CC BY 3.0, https://commons.wikimedia.org/w/index.php?curid=21536441.

Bekoff holds that his field, cognitive ethology, depends upon these tools, analogy, anecdote and anthropomorphism, to interpret their findings into a language that makes it possible for us to grasp. "As humans who study other animals, we can only describe and explain their behaviour using words with which we're familiar from a human-centred point of view" (Bekoff 2007, 123). What we can take from this is that anthropomorphism is a linguistic tool that makes the thoughts and feelings of other animals accessible to us—and as such is, as Bekoff points out, both inevitable and also involuntary (Bekoff 2007, 123). One poignant example given by Bekoff considers a young fox (Figure 2.1).

> How do we know a young red fox misses his mother? He squeals, searches for her, and when the two are reunited, he snuggles her closely, closes his eyes, and quietly falls asleep. We could restrict our description to just [...] the fox's actions and manner, but it is [...] the mother's absence that help clarify that [...] the squealing and snuggling didn't indicate [...] some other behaviour. (Bekoff 2007, 121)

Anthropomorphisms in the past

From an archaeological point of view, it is significant that the human bent for anthropomorphizing probably has great time depth. Literature serves as a lucid demonstration of this throughout history: poets and storytellers have created myths and tales about animals in which the animals are imbued with a rich inner life of emotions and thoughts. By going back in time from recent times to ancient Greece, I will give some examples. Rainer Maria Rilke's poem *Der Panther*, composed in Paris in 1902 serves as an eloquent example closer to our time:

His gaze against the sweeping of the bars
has grown so weary, it can hold no more.
To him, there seem to be a thousand bars
and back behind those thousand bars no world.
The soft the supple step and sturdy pace,
that in the smallest of all circles turns,
moves like a dance of strength around a core
in which a mighty will is standing stunned.
Only at times the pupil's curtain slides
up soundlessly—. An image enters then,
goes through the tensioned stillness of the limbs –
and in the heart ceases to be.[4]

Rilke is clearly anthropomorphizing here, yet his poem is based upon an encounter with a real life panther behind bars in Paris Jardins des Plantes. Could it be that he truly, empathically felt what the panther was feeling? Or was he projecting his own expectations—loss of dignity, boredom, sadness, etc—onto the caged animal?

Reaching further back in time, the evocative Norse Eddic poems give several examples of perceptions of the inner lives of animals. A poignant example is the death of Sigurd, the dragonslayer, described in "The second lay of Guðrún" and "Fragment of a Sigurd Lay." In "The second lay of Guðrún" verses 4 and 5, Guðrún recounts how it happened that she found out that her husband Sigurd was dead:

4. From the Thing ran Grani with thundering feet,
But thence did Sigurth himself come never;
Covered with sweat was the saddle-bearer,
Wont the warrior's weight to bear.

Guðrún held the rein of the horse and began to cry, as she understood what had happened:

5. Weeping I sought with Grani to speak,
With tear-wet cheeks[5] for the tale I asked;
The head of Grani was bowed to the grass,
The steed knew well his master was slain. (The Poetic Edda 1936b)

Also in the poem "Fragment of a Sigurd Lay," the following line in verse 7 refers to Grani mourning Sigurd: "The gray horse mourns by his master dead" (The Poetic Edda 1936a).

The story demonstrates that human–horse bonds were deemed possible, and that horses were not only seen as sentient, but also held the potential agency to both have and express strong feelings. Further, I find it remarkable that Grani in this story ran to Guðrún and grieved in front of her. This reveals a belief in horses having the ability to understand human relationships; Grani seeks out the wife of Sigurd to mourn together with her, and to thus facilitate a mutual recognition of the love they both bore for Sig-

4. English translation by Stanley Appelbaum. https://en.wikipedia.org/wiki/The_Panther_(poem)

5. In the original Norse Guðrúnarkviða II, it is ambigous which of the two has tear-wet cheeks. Even though the English translator has chosen to let Guðrún's cheeks be covered in tears, two Norwegian translators understand the text as though that Grani has tears in his eyes.

urd—a recognition in which they can find solace, as two persons who loved this man deeply and miss him dearly (Armstrong Oma 2016b). A similar argument has been proposed by Gala Argent (2010, 162) suggesting that "horses can and do bond deeply with specific people; horses, too, can choose particular people as their 'person of a lifetime'," exemplified by the account of Alexander the Great and his renowned horse Bucephalos.

The telling of Grani's reaction could be seen as another example of anthropomorphism. Guðrún mourns with the horse, and she might project her own feelings onto Grani. The real truth is beyond reach, but clearly, this points to a perception of an emotional component to the human–horse relationship in the Scandinavian Iron Age, considering that the mourners have placed horses—or tokens of them—with their dead (Armstrong Oma 2011, 2016b; Oma 2000, 2001, 2004).

Another example of anthropomorphizing storytelling with a great time depth is Aesop's fables, in which animals are the main characters. The fables go back to around 500 BCE and were written down around 300 BCE. Amongst the fables, there is a well-known story of a fox:

> A hungry fox saw some fine bunches of grapes hanging from a vine that was trained along a high trellis, and did his best to reach them by jumping as high as he could into the air. But it was all in vain, for they were just out of reach: so he gave up trying, and walked away with an air of dignity and unconcern, remarking, "I thought those grapes were ripe, but I see now they are quite sour."

Without doubt this fable anthropomorphizes the fox, when he fails to achieve his mission he pretends to never have wanted the grapes in the first place, being too proud to admit defeat. This means that there was a tradition for anthropomorphizing—or egomorphizing—more than 2000 years ago—and no doubt this extends back in time.

Another interesting example are birds of prey that belong to the family of corvids, crows and ravens. These birds are not popular in present-day Norway, and there are segments of society bent upon their extermination. However, in pagan Norse mythology, Huginn and Muninn were the ravens of Odin, the king of the gods. These two ravens were perched on Odin's shoulders, and their names indicate that they represented thought (Huginn) and memory (Muninn). They were guardians of esoteric knowledge and magic, and from their elevated position on the shoulders of the god, Odin the All-father, they could see the whole world (see Figure 2.2; Hedeager 2011).

Recently, there has been an upsurge of research in Animal Studies on the capacities of corvids, and they are demonstrably unusually clever with mental capacities that match those of primates. They can make tools and have independent problem solving skills, advanced communication skills—and they play for the fun of it (Marzluff and Angell 2012). This kind of knowledge and understanding of corvids must at some level have been understood by the ancient Norse, who worshipped these birds for the very qualities that have been uncovered and described by Animal Studies. To have such an extraordinary degree of knowledge of other kinds of beings in the world shows that people in the past observed the world around them acutely and accurately with far greater interest, reflection and awareness than most people today, and they knew the capabilities of the animals that inhabited their world intimately. Odin's ravens tell us that in the Norse Iron Age, corvids were valued for their intelligence and ability to com-

Figure 2.2 Odin on Hlidskjalv, with Huginn and Muninn perched on the armrests? Found at Lejre in Denmark. By Mogens Engelund—Own work, CC BY-SA 3.0, https://commons.wikimedia.org/w/index.php?curid=8541916

municate. Being airborne, they had a wide reach, and they lent their unique capacities to the god who could not be Allfather without their aid.

A discussion of anthropomorphism holds particular importance to archaeology, since we cannot access past human-animal relationships directly, but can only surmize from the objects and fragments—and their contexts—that can be observed as relics of practice. The archaeological situation gives the scope to make inferences about how situations grew forth from particular co-habitations and their inherent choices, and therefore—how humans *perceived* animals. Because of this particular situation, it is imperative to have a clear understanding of how human perception of animals can be formed.

As an archaeologist, the question that stands out is—would the concept of anthropomorphism have made any sense to people in those societies? The examples referred to here are from written sources, and reveal some structuring principles of the societies from which they have sprung. Clearly, those that forged these stories considered animals to be sentient beings. Further, it appears that in many cases people in the past didn't necessarily perceive a difference in kind between themselves and animals, therefore anthropomorphism as a concept would have made little sense to them—probably the idea of egomorphism would have resonated much more deeply. The same is true for many indigenous societies (Morris 1998, 2000; Willerslev 2007).

Mirror neurons

Is there, then, any truth to our anthropomorphizing interpretations of animals? One indicator that suggests that we do indeed get the right end of the stick with our anthropomorphizing is the discovery of a group of cells in the brain called mirror neurons. These cells were first discovered in macaque monkeys, and the cells fire both when the individual performs an action, but also when the individual observes others performing actions or displaying emotions (Rizzolatti *et al.* 2009). The neuron cells in these monkeys reacted both when other monkeys and humans performed actions. This changes our scientific understanding of how we understand, and respond to, the actions of others. Previously, it was assumed that we cognitively interpreted the behaviour of others, based upon inferential reasoning. The discovery of mirror neurons has demonstrated that there is a much more direct link between the experience of our own behaviour and that of others.

> However, the ease with which we usually understand what others are doing suggests that an alternative mechanism might be involved in action perception. The essence of this alternative system is that actions performed by others, after being processed in the visual system, are directly mapped onto observers' motor representations of the same actions. The observers are aware of the outcomes of their own actions, so the occurrence of a neural pattern similar to that present during their own voluntary motor acts will enable them to understand the actions of others. (Rizzolatti *et al.* 2009, 24)

Without question, this discovery has radically changed—and will continue to change—the way individuals understand actions, intentions and emotions. In identifying this mechanism, a profound impact upon disciplines ranging from cognitive neurosciences to sociology and philosophy will be seen (Rizzolatti *et al.* 2009, 32–33).

Some researchers have suggested that mirror neurons go a long way towards explaining the phenomenon of empathy. Findings suggest that feeling emotions and recognizing them in others are mediated by the same neural substrate—meaning that we directly experience the emotions of others (Rizzolatti *et al.* 2009, 27).

Another consequence of mirror neurons is that it is not just humans who directly experience the behaviour of animals—and thus engage in the act of anthropomorphizing. The experiment with the macaque monkeys demonstrates that beyond a doubt, other animals also experience each other's behaviour. Thus, it is fair to suggest that not only do humans anthropomorphize—but animals zoomorphize human behaviour, too. This gives a further dimension to all those "dotty" humans who claim that nobody understand them as well as their companion animals.

However, mirror neurons are not a stand-in for strong intersubjective bonds with time-depth that spring from knowing each other deeply—such a knowledge of others arises out of the act of relating. It is therefore not a substitute for relationships, with all that they entail. What mirror neurons do is to enable different species to know and understand each other based upon their own experiences of being in the world.

Difference in degree or in kind?

According to mirror neurons, it might in fact be more accurate to refer to our understanding of the behaviour of other beings as egomorphism (cf. Milton 2005), since it is based upon our direct experience of the behaviour of others, rather than us cognitively interpreting behaviour. We, humans, recognize ourselves as individuals in individual animals, and they recognize themselves in each other, and in us. We do not stand face to face with animals as abstracted "us"—"they," but as individual to individual, regardless of species. Encounters can be tangential or repetitive, but still the fundamental autonomy of the other is recognized. However, this statement should not be confused with intersubjective relationality.

Mirror neurons also addresses one of the main pitfalls of anthropocentrism, namely the reduction of animals to lesser versions of ourselves and to attribute feelings and notions to them that they might not have. However, not all interpretations of animals are experiential and based upon mirror neurons. One example is horse owners that wrongly think, based upon themselves feeling cold, that their horses get cold when they are outside and cover them in blankets when horses in fact have a thick fur and are well equipped to deal with all sorts of weather.[6] Anthropomorphism can in many situations be harmful to animals. Irvine (2004, 75) points out that "anthropomorphic projection can do considerable harm when it produces an image of what the animal must 'really' be like." Haraway warns us of this in her Companion Species Manifesto (2003) when she reminds us that dogs are not furry babies. They are their own creatures, and as such, mysteriously different from us. What this difference consists of is not always easy to tease out. Bekoff (2002, 10) also holds that it is "simplistic and misleading to assume that 'they' are merely 'us' in disguise." But where does the difference lie? Darwin "emphasized mental continuity across species and thus attributed cognitive states to non-human animals" (Irvine 2004, 63), a perspective referred to as "anecdotal cognitivism" (Jamieson and Bekoff 1993). Famously, Darwin (1936 [1871], 448) stated that the difference between humans and animals is one of degree rather than kind, "there is no fundamental difference between man and the higher mammals in their mental faculties." Some 150 years later, the scientists who formulated the "Cambridge declaration of consciousness" held the scientific evidence to support Darwin's claim. And they went further, and stated that this is true not only for the "higher mammals," but also, for example, for octopuses.

But, the continuity between "us and them" that Darwin advocated stands in stark contrast to most of the Western cultural history, both regarding the fundamentals from the monotheistic religion and much of the Western philosophical tradition (McFarland and Hediger 2009, 4). When approaching human–animal relations in prehistory, several colleagues (Argent 2011; Brittain and Overton 2013; Johannsen 2012; Mlekuz 2012) have advocated a view that precludes Cartesian-informed humanism, and my own research supports such a view. Therefore, I give primacy to Darwin's claim of continuity between human and non-human animals. This is also supported by a study which sought to establish comparative perceptions of the mental capacities of humans and animals, and concluded that "the dog and boy were seen as *quantitatively different*, but *qualitatively similar*" (Rasmussen and Rajecki 1995, original emphasis). The dog and boy were thought to have the same kinds of mentality, but the boy was thought to have a greater mental capacity (Milton 2005, 257).

6. Personal communication Lynda Birke 15.11.2011

Perceiving animals as subjects or objects

Another point that needs to be addressed when considering human–animal relations in the past is what kind of status animals can be afforded—are they given subject status, or have they been construed as objects? Being humans, we interpret animals, and their roles—as discussed above, we anthropomorphize—or egomorphize. Fundamentally, humans choose to look at animals as one or the other way: subject or object—even though, as Buller (2012) points out, animals flirt in-between subject and object (see also Orton 2010).

Traditionally, scientists have looked at animals as objects—simply because that is what they are trained to do (Bekoff 2007, 114). As outlined in chapter 1—archaeology, and in particular zooarchaeology, has traditionally followed this lead, striving to give the discipline legitimacy by adopting scientific means and methods. This striving for objectivity in researchers has led to a neglect of the subjective position of the researcher and to "a view from nowhere" (Nagel 1986), that does not rely on a "particular vantage point and is as impersonal as it is possible to be [...] *distancing away from* engagement, to gain what is assumed to be an impartial, 'perspective-less' view." (Wemelsfelder 2012, 224, original emphasis). Wemelsfelder holds that this "perspective-less" position is problematic when studying sentient others. Further, feminists such as Sandra Harding (1986) and Donna Haraway (1988) have, by way of Standpoint Theory, uncloaked the "view from nowhere" as a "god-trick." This means that everybody has a "view from somewhere," embedded in the *zeitgeist* of their time, and socio-cultural biases are always and everywhere present.

Dimitrij Mlekuz (2012) rightly points out that objectifying animals leads to a vertical ontology with humans on top that created a hegemonic subject-object dichotomy. Archaeologists have followed this lead, and taken the cue from the "real" scientists. However, animals are by their nature subjective beings (see Figure 2.3). I follow Marc Bekoff (2007, 120)—and many others—that state that animals don't merely act "as if" they have feelings, they do indeed have the ability to feel. In fact, trying to explain away this fact, by using so-called parsimonious explanations, in which behaviour is explained as a chain response of mechanistic input and response by instinct, has led science into a quagmire. This narrative form is frequently termed the externalistic approach, and is, for example, adopted by scientists that label animals involved in experiments as "data" (Birke *et al.* 2004, 171). This approach is firmly grounded in the Cartesian separation, based upon Descartes' (1912 [1637], 116) notion of animals as biological automata. In biology and ethology the trend has been to explain animal behaviour as parsimonious and mechanistic. These two concepts are interrelated, since the mechanistic model accounts for the parsimony of explanation and, similarly, the parsimonious explanation is the result of the mechanisms. According to this approach, animals are pre-programmed to perform specific patterns of behaviour without any ability of awareness or consciousness, acting like automata (Midgley 1988, 38–40).

A pendulum swings both ways, and a contrasting approach is referred to as the internalistic approach which often embraces an extreme anthropomorphism. This approach reflects, for example, people's attitudes towards pets (e.g. Edelman 2005; Serpell 1996 [1986]; Swabe 2005), where animals are treated as though they were humans. Within this approach, animals are considered to have varying degrees of intelligence and

Figure 2.3 Human-animal relationships on the farm are often slippery when it comes to how an animal is understood. Is this farmer meeting subjects or objects? Photograph by Anne Helen Robberstad / Jærmuseet.

awareness, and most pet owners will claim that the pet understands what s/he is told, and the pet's behaviour is read as though s/he is driven by the same motivations as those of humans (Edelman 2005, 101; Swabe 2005). In this approach, animals are frequently sentimentalized, similar to children (Arluke and Sanders 1996; Callicott 1989, 49–59; Forrest *et al.* 2005; Knight 2005; Sanders 1993; Sofaer Derevenski 2000; Swabe 2005, 104; Theodossopoulos 2005, 22). This approach frequently treats animals as quasi humans (see Swabe 2005 for in-depth discussion), an artificial projection of humanness onto animals. However, I argue that this approach can also fundamentally be objectivizing, as it does not consider animals as themselves, but as some*thing* according to the eyes of the beholder. H. Peter Steeves (1999, 9) puts it thus: "sentimentalism and paternalism are forms of violence. They reduce the animal, speak (inappropriately) for the animal, and fail to do justice to our being together" (with reference to Dallery 1999 and Laycock 1999).

What is it, then, to afford animals subject status? Haraway puts it thus:

> contrary to a lot of dangerous and unethical projection in the Western world that makes domestic canines into furry children, dogs are not about oneself. Indeed, that is the beauty of dogs. They are not a projection, nor the realization of an intention, nor the telos of anything. They are dogs, i.e...., a species in obligatory, constitutive, protean relationship with human beings. (2003, 11–12; 2004, 300)

When Haraway's clarification is heeded, subject status can emerge.

The externalistic and internalistic approaches reveal a rift in how animals are not only perceived, but also how they are treated. Franklin (1999) points out that in contemporary Western culture some animals can be sentimentalized and cherished, as long as others can be produced in intensive factory conditions for nutritional needs (see also Noske 1997). The use of language in these two forms of narratives reveals what awkward positions animals are subjected to in the modern world. This is comparable to the gap in language in the different strands of the archaeological discourse that deals with animals, the language of the internalistic approach lies closer to how narratives including animals are recounted when religion, ritual and cult are at the heart of the matter, whereas animals are represented in an externalistic, parsimoneous way by zooarchaeologists dealing with subsistence and economy (Armstrong Oma 2007, 59–60).

In our current culture, we could argue that pets are dominantly seen as subjects, whereas livestock on factory farms are objectified. A major difference between those two perspectives is whether or not animals are seen as beings with agential powers—and these attitudes stem from the situatedness of the perceiver. The difference, therefore, lies not only in the eyes of the perceiver, but also in the context. Situations that allow seeing animals as subjects can be dictated by a variety of factors. One of these, which is tangible and can also be found in the archaeological record, is the scale of distance versus proximity. Closeness allows for a degree of observation that distance will not. Seeing is knowing, and knowing is to develop a degree of understanding of the other. This is true of pets, farm animals and even the prey of hunters.

To start with the latter, many ethnographic accounts give descriptions of how the hunter seeks to know the animals s/he is hunting. Anthropologist Rane Willerslev lived with the hunters of the Yukaghirs and has portrayed how they, through both close observation and metaphysical communications (dreams and visions) sought to know, befriend and engage with the moose they were hunting. Willerslev (2007, 2012) describes how the Yukaghirs saw themselves and the moose as living in a world of mirrors, in which both parties reflected aspects of each other (see also McNiven 2010). There is an intense engagement between hunter and prey, and much lore, possibly superstition, of how the animal "gives" itself to the hunter (Nadasdy 2007).

Regarding farm animals, they are frequently objectified, both in factory farms and in the consumer's mentality. Little is known of the lives of those beings that end up on the dinner plate. This implies a large degree of distancing between consumers and farm animals. But the distancing is also found in the factory farm, where modern technology now allows for a larger distance between the animals and the stock-persons. One example of this is that of the recent development of the "milking robot" (Bear *et al.* 2012). Whereas previously, no more than 60–70 years ago,[7] cows were milked by hand (as indeed they have been since the secondary product revolution), technology has increasingly removed the farmer from intimacy with her/his livestock. This development has, along with the industrial revolution, contributed to the increasing distancing between people and farm animals (Bear *et al.* 2012). As a result, farm animals are in most instances regarded as objects, rather than as subjects with an ability to act upon

7. Milking machines were introduced in the 1920s and were in common use by the 1950s. Personal communication: Tarald Magne Oma, 30.01.2014.

the world. Such are the chains that bind them that they are very seldom allowed the space to express any kind of individual personality.

However, not all modern farmers adhere to this mind-set. In her book, *The Secret Life of Cows*, farmer Rosamund Young (2003) discusses the social life of the cows and other domestic animals on her farm. She has observed the individual animals closely, and noted that they form friendships that last throughout their lives. If a close friend is away for a while (for example, kept inside the barn due to illness), the cow left out in the meadow will keep an eye out for her friend, and display a great show of affection once they are reunited, also following lengthy separation.

Loving animals

In line with the observations made by Young above, a number of ethologists have reported that animals display behaviour that seems to be emotional. For example, mammals and birds seem to fall in love and often form partnerships for life (e.g. Bekoff 2002, 20–21), in fact, Bekoff (2002, 20–21) claims that 90% of all bird species are monogamous. Konrad Lorenz and his colleagues (Lorenz *et al.* 1991) observed that when the greylag geese fall in love (sic! According to Lorenz, they do really fall in love), their behaviour closely resembles that of their human counterparts. From the perspective of Bronze Age farmers, these are interesting findings. The farmers of the past would have observed similar kinds of behaviours.

Some people might doubt that animals have the ability to feel love. And even if they do, what relevance does it have to the study of animals in past societies? Bekoff (2002, 20) holds that it is "unlikely that love appears in humans with no evolutionary precursor." This observation does, of course, imply that if contemporary animals display behaviours that we, based upon our own experiences, classify as emotional—such as love, then people in the past in all likelihood made similar observations. This brings us back to the question of anthropocentrism and egocentrism,—and to what degree these concepts are valid for the past. Considering that, as previously suggested, in most societies in prehistory the distance between humans and animals was considerably less than for the majority of people today, people in the past would have had ample opportunity to study animal behaviour and draw their own conclusions about it. Bekoff (2002, 21) suggests that whereas human love can be confusing (presumably due to the mystifying nature of language), animal love is less so, because animals do not filter emotions like we do. In contrast, their love is out there for all to feel—and he accentuates: that is, for those who *choose* to feel it. I would claim that most people who have chosen to share their lives with animals indeed have chosen to feel that love.

Observations and interpretations aside, there now exists scientific evidence that concludes that humans and animals share neurochemicals and biochemistry that is linked to emotions (Bekoff 2002, 17). It is well known that hormones play a large role when it comes to social bonding between members of the same species, but hormonal biochemistry actually crosses the species boundary to facilitate social bonding between different species. The hormone oxytocin is often dubbed the "love hormone" and is connected with attachment and social bonding. It is in particular associated with breast-feeding, since it is released into the bodies of lactating women. It also prepares women for birth.

This hormone is not exclusive to humans, all mammals have oxytocin as a component in their biochemistry (Heinrichs and Domes 2008). However, oxytocin also works across the species boundary. For example, studies have shown that when a human strokes a dog, the hormone oxytocin is released into the body of not only the human doing the stroking, but also the dog that is stroked (Handlin 2010). Similarly, oxytocin is released into the body when a dog looks at her/his guardian (Nagasawa *et al.* 2009). This is likely not a recent phenomenon. Mammal biochemistry has evolved over eons, and oxytocin has been a neurochemical component in our biochemistry throughout the ages. Inciting love for her offspring in a mother is an obvious evolutionary advantage, thus we can safely assume that oxytocin has great time depth and goes back to the development of the mammal species. When we study human–animal interactions in the past, we must therefore consider the factor that forming relationships to animals could potentially simply make both the people and the animals feel good.

Epistemologically, this research is interesting, as its language balances in-between hard and soft sciences. The use of language in the archaeological discourse of Bronze Age animals tends to be either/or, with a lack of overlap in language when it comes to subsistence and animals and animals and the divine (Armstrong Oma 2008a, 2012, 2013a). However, as such studies win ground in "hard science," archaeology might be coaxed to follow suit.

Towards an archaeology of animals as subjects and agents

The rationale of this chapter is to illuminate the need to reconsider the tenets underlying the theoretical framework by which archaeology functions in relation to animals. Simply "add animals and stir" is not going to cut it (Argent 2011). Rather, we need to reconsider the framework we are embedded in, and shed what feminist scholar Donna Haraway describes as the "culturally normal fantasy of human exceptionalism... the premise that humanity alone is not a spatial and temporal web of interspecies dependencies" (2008, 11). The next chapter goes further in setting up an alternative framework in which animal agency is firmly established. Novel directions in philosophy are drawn upon, such as feminism and posthumanism; these form a stark contrast to the humanist mindset.

— 3 —

Animal Agency

Challenging the species barriers

This chapter builds upon the discussions in chapter 2, since a careful consideration of sentience and perceptions of it is a fundamental starting point to explore animal agency. As outlined in the previous chapter, in the "hard" sciences, such as neuroscience, biology and ethology, there is a growing recognition that all kinds of beings that fall into the animal category are beings with agentive abilities. In the last couple of years there has been an enormous growth in animal studies, and new reports of the amazing abilities of animals pour out on a daily basis. These include parrots that can distinguish colours, shapes, can spell and do maths (e.g. Pepperberg 2000), corvids that use tools and whose intelligence seems to be converged with primates (Seed *et al.* 2009), and primates that can do advanced maths and recognize and remember numbers at a speed that no human can hope to achieve (for example Martin *et al.* 2011). Youtube is overflowing with videos of animals that perform amazing—and highly entertaining—tasks. Researchers within biology and ethology have discovered that tool use, language use, consciousness of self, culture, art and rationality are capacities and qualities that animals also have. Such features can no longer be reliably used to draw up species barriers that separate humans and animals (Bekoff 2002, 13). More and more, these studies form a complex image of how animals are like us, how they have intelligence like us, and in some cases even exceeding our own. But unfortunately, the questions most frequently asked by such researchers are "how are non-human animals like humans?" Really interesting answers surface when scholars of Animal Studies ask "how are animals brilliant as themselves?"

Animal agency

If there is no essentialist difference between human and non-human animals, then the core elements that comprise the fundamentals of being human is also true for being animal. One of these is agency. Agency is, according to the Merriam-Webster dictionary, "the capacity or state of acting," and more specifically I would say it is to act from a place of awareness and integrity. My position is that the ability to act, to move into action, gives beings agency by their actions, having an effect and leaving tangible imprints upon the world.

Animals and their agency should hold particular interest to archaeology, since human and non-human animals share a deep history not only concerning predator/prey relations, but also as co-creators of the world. This pertains to the development over time of human–animal interactions in the past and present; both in terms of their relationships with humans, and also human perception of animals' nature and agency. Archaeology has a role to play in this respect, as the discipline provides an essential time-depth to understanding current cultures and societies that are historically situated within particular trajectories (see also O'Connor 2013). The processes whereby joint human and animal societies have been formed have not been created in a vacuum. Understanding these processes gives important insights into both specific questions like domestication and art, and even to the rise and fall of civilizations.

The archaeological discourse on animals has traditionally been devoid of the potential agency of animals, but rather focused on animals as either containers for human symbolism, as economic strategies, or to segment them into abstract categories of inert variables (Armstrong Oma and Birke 2013). As in the humanities at large, human agency has been granted supremacy. This attitude has seeped into archaeology by way of the Enlightenment roots of the humanities, as discussed in chapter 1. In a philosophical sense this entails that ontologically the nature of being is the nature of human being; the nature of action is of human action (Johannsen 2012, 305). But animals are more than cultural abstractions: what is lacking is considerations of the animals *as themselves*. However, there currently is, within Animal Studies and Human-Animal Studies, a growing interdisciplinary recognition that many animals possess characteristics such as intelligence, emotion and awareness that vary from humans by degree rather than kind (DeMello 2012; Shapiro and DeMello 2010). Animals are alive, active participants in their worlds, and the spaces where those worlds intersect and enmesh with humans are often messy and difficult to divide into clean compartments. In addition to how humans "use" them, animals often take part in subjectified relationships with humans that impact both species at various levels of scale. For example, Birke, Bryld and Lykke (2004, 172–173) describe how rats in labs defy their definition as "data" for animal experiments by requiring that humans facilitate their biological and social needs and perform accordingly.

The human–animal hyphen

The theoretical foundation of my project is a perspective in which animals are active co-creators of the world (Haraway 2003, 2004). This has a particular relevance for farming societies. By way of their sentience, animals possess agency by their ability to purposefully act upon the world, unless severely physically restrained. On the farm, space is created, shared and mutually constituted by humans and domestic animals. Meanings arise and practices are constituted as joint actions unfold; whilst herding, milking, plucking wool, walking together, resting together, and creating spaces and thresholds (Armstrong Oma 2007). The consideration and care in which individuals are allowed to carve out their personal place is created everywhere on the farm.

Fundamental to this argument is that the relationship between living animals and humans is not a one-way communication; rather it is a process of becoming between individual humans and animals. This perspective creates an understanding of animals as a doing or a becoming, formed by social interactions (Birke *et al.* 2004, 169). The becoming emerges from mutual human–animal engagements that are born from the co-habiting situation of sharing the world—i.e. on the farm, as humans and animals alike dwell there. Considering that animals are sentient beings, it is vital to recognize that animals are in a position to engage in what Birke, Bryld and Lykke (2004) term "intra-action" (see also Barad 2007), which denotes that humans and animals are engaged in mutual decision-making, steeped in a choreographed and performed co-creation of behaviour (Birke *et al.* 2004, 174). A field of intra-agency, at the conjunction between humans and animals, emerges from all of these joint actions. I have previously argued that intra-action takes place in spatial locales, so-called meeting points (Armstrong Oma 2007, 2013b).

The density of the language I use to talk about intra-agency is borne from the complexity of the situation, and the fact that intra-agency is not like Hegelian dialectics. The key word is *emergence*, relationships between humans and animals emerge from the way they see, sense, relate, act, behave, accept, reject, and so on, each other. The intra-agency emerges from the hyphen, human—animal (e.g. Brittain and Overton 2013). The emergence creates something that is larger than the sum of the parts, new kinds of agents (see also Westling 2014). This perspective on relationships is in line with Ingold's (2011) idea of meshwork, in which the lines in-between the points are what matters as they are the site of becoming. For example, Game (2001) describes horse and rider as rider being part horse, and horse being part rider, thus a creature beyond horse and human is born: the centaur (see Figure 3.1).

The act of relating, and thus the relationship, is formed in the encounter between humans and animals, and it is by mutual action, becoming, and performing that a field of intra-action emerges. Thus, when humans and animals face each other they become Others to each other (Cudworth 2011) since the impact of the encounter upon both becomes fundamental to their understanding of themselves. This works as a discursive process that takes place in human–animal encounters, so the spatial locations where humans and animals meet are arenas where intra-action materializes.

Such an understanding of animals, and their relationship to humans, is more inclusive and thereby more useful when examining the role of animals in the lives of humans. It also caters for societies that practice husbandry, where a close proximity between living animals and humans take place. The repetitive nature of the field of action and the associated mutual becoming has implications for the social lives of agents in farming societies. For example, baby animals must socialize themselves in relation to humans and non-human animals both.

A brief word on companion species is needed before proceeding. Haraway (2008) invented the term companion species, and uses it to talk about myriad species, mess-mates, being at table with all kinds of beings, such as bacteria etc. Even though I have great sympathy for her project, this does not resonate with my approach. Although I do contend that animals and humans are companion species to each other, I am interested in the kinds of becomings that emerge from specific and historically situated happenings that are physical meetings between beings that recognize each other's presence.

Figure 3.1 Sandro Botticelli's Pallas and the centaur. The Yorck Project: 10.000 Meisterwerke der Malerei. DVD-ROM, 2002. ISBN 3936122202. Distributed by DIRECTMEDIA Publishing GmbH.

Agency within and outside the archaeological discourse

In archaeology, agency is a term that, one could argue, has been abused by overuse, particularly because the perceived agency of the material world has merged with the theoretical discourse on materiality (examples include DeMarrais *et al.* 2004; Miller 2005; Olsen 2003; Renfrew 2004). As such, archaeology justly falls under the critique by Tim Ingold (2011, 28): "the material world can only be brought back to life in the dreams of theorists by conjuring a magical mind-dust"—namely agency. Further, even though much theorizing has been eager to attribute agency to material objects, animals have curiously been left out (Johannsen 2012)—even though animals by their ability to act *by* and *as* themselves fit the bill. Ingold is also strangely silent when it comes to this question, albeit he discusses the ludicrous (sic! In his words) idea of attributing agency to objects at length. When the ANT (actor-network theory) meets his SPIDER (Skilled Practice Involves Developmentally Embodied Responsiveness), he warns us of confusing agency with intelligence (2011, 93–94). This again goes to show that the nature of animal agency is not clear-cut, and nor, for that matter, is human agency. Agency is a slippery concept, and as pointed out by Noortje Marres (2005), this "ungraspability" could be an essential aspect of agency. The debate as to which factors should be present to have agency ranges from intentionality, free will and morality. New research in biology and ethology is constantly challenging the common perception of such qualities as human-only and bridges the gap between humans and animals (e.g. Bekoff 2002, 2007; Wemelsfelder 2001, 2012; Wemelsfelder *et al.* 2001). This research is instrumental to countering Ingold's claim of agency as just "magical mind-dust"—at least concerning animal agency. In the introduction to the volume *Animals and Agency*, the editors McFarland and Hediger (2009, 8) puts it thus: "no matter how we consider agency, the result is that either humans are more like the other animals or the other animals are more like humans than we have comfortably thought in the past." I would add, than we have thought since the advent of humanism.

John Law and Annemarie Mol (2010, 58) explores the agency of a Cumbrian sheep situated in time-space (2001 in England in the middle of the foot-and-mouth epidemic). Their starting point is material semiotics and actor-network theory (ANT), which claims that entities give each other being, that they enact each other. Agency is thus "ubiquitous, endlessly extended through webs of materialized relations." This means, as Law and Mol go on to say, that "an actor does not act alone. It acts in relation to other actors, linked up with them. This means that it is also always being acted upon. To act is not to master, for the results of what is being done are often unexpected." This perspective slips away from the distinction between mastery and being-mastered (Law and Mol 2010, 57). Beings act and are enacted. However, actors are not agents—the kind of agency that is discussed in relation to the Cumbrian sheep gives primacy to all nodes within a network. I find that this framework is useful to break up dichotomies, such as power discourses, which is demonstrated by their discussion of sheep. For example, commenting on a statement given by a farmer:

> If a sheep is grabbed and turned on its back then it is being done to, but even a sheep brutalized in this way is not passive. To begin with, it is *surprised*. And, let us face it, it also *allows* the human beings around it to act in this brutal way. It does not, for example, respond by biting them [...] What is more: this sheep is active in remembering its man-handling. It may have been a pet lamb last year, a lamb fed with bottles that learned to trust human beings, but next time the farmer enters the shed, it may well *decide* to skip the pleasantries. It is not stupid! (Law and Mol 2010, 67, original emphasis)

Indeed! So, sheep enacted also act, and I would argue that the sheep is an agent whether s/he decides to skip the pleasantries or not.

Animals as "expressive behavers"

A poignant question is, how do we understand and talk about animal agency? Françoise Wemelsfelder has devoted much research to this question, seeking to develop "scientific approaches for the study of animals as whole sentient beings (i.e. as subjects rather than objects), bringing insights from philosophy of mind into the study of animal emotion"[1] and "a practical methodology for welfare assessment that would enable scientists to work directly with animals as sentient beings" (Wemelsfelder 2012, 225). Her research has used a method known as "free choice profiling approach" as a qualitative assessment of the whole animal (Wemelsfelder *et al.* 2001). Basically, the method records the spontaneous reactions of observers, in which the observers have complete freedom to generate their own descriptive terminology (Wemelsfelder *et al.* 2001, 210). Her approach addresses anthropomorphism from a broad spectrum, comparing the assessments made by a variety of observers. These showed a significant degree of consistency, demonstrating "the strong evidence of the internal validity of spontaneous qualitative assessments" (see Rousing and Wemelsfelder 2006; Wemelsfelder *et al.* 2001, 218). What is striking about the results is that animals emerge as "expressive 'behavers' rather than assembled strings of 'behaviours'" (Wemelsfelder *et al.* 2001, 219). Citing Nagel (1986), Goodwin (1999) and Sheets-Johnstone (1999), the explanation given by Wemelsfelder *et al.* (2001) for animals as expressive behavers is their inherent agency, and thus the descriptions of the behavioural expressions of animals is a description of their agency, reflecting the whole animal as a dynamic focus. Wemelsfelder (2001) argues that by looking at the whole animal the study of animal experience should gain empirical ground, and become less, not more, vulnerable at the risk of anthropomorphic projection.

Going back to archaeology, free choice profiling is an interesting and also useful analogy. I suggest that it is also the approach that people in past societies used when faced with other species. Rather than fragmenting the animal and singling out one or two behavioural traits, the whole animal would be appraised, and interpretations were based on as full an experience as possible—seeing animals as "expressive behavers." For the animals to be experienced as such, their agency was a prerequisite.

Henry Buller states "moving and being are one" (2012, 139), and suggests that the movement of our bodies is the common denominator in human and non-human animal agency, since gestures and bodily responses are shared by humans and animals (Figure 3.2). As such, humans and animals both are "expressive behavers." In his article from 2012 Buller quotes the poet Ruth Padel and her poem "Tiger drinking at forest pool" (2004) in which he finds the line "one slash of light, then gone" particularly evocative, where the animal by its movement is "flirting" between worlds, and is just beyond a purely scientific and a phenomenological understanding, between "object" or "subject" (2012, 139). Moreover, animals make themselves known to us through their movements, and the spaces that they trace and occupy invite our orderings (Philo and Wilbert 2000). Buller wishes to "explore the idea of studying, observing, practicing and sharing embodied movement as a means to experiment with a new mode of connectivity and "being" with non-human animals" (2012, 141). His quest is along the same line as my project, in which I seek to explore the effect of such movements, and their inherent agency, on farming societies in the Bronze Age. Interesting is how animals by their agency and volatile movements transgress in space, and either intrude or flee; again—they flirt between the worlds of lost and found.

1. http://www.sruc.ac.uk/fwemelsfelder

Figure 3.2 "One slash of light." Sheepdog chasing sticks. Photograph by Kristin Armstrong Oma.

What kind of agency is animal agency?

Another, equally important question, is posed by Erika Cudworth (2011, 17–18), "What agency might other species—even relatively privileged ones—have, within systemic relations of human domination?" Cudworth puts her finger on the asymmetrical nature of human-animal relationships that is overwhelmingly seen in for example farming, and power relations are also definitely a factor to be reckoned with in past societies. On a farm, animals should not—even when their agency is admitted, and status as producers is granted—be confused with humans. Their status and the place from which they acted were different to that of humans. Their status differs ontologically from humans—and further from species to species.

I hold that there is no such thing as a perfect symmetry in any relation, be it between partners, parents and children, or different human-animal relationships. Extensive studies in biology have repeatedly demonstrated that the same is true for different species of animals—relations are organized in some sort of hierarchy, however flexible this might be. In a debate that involves Marc Bekoff, L. David Mech and Lee Charles Kelley, Bekoff (2012a), states that "animal behavior experts who study a wide variety of species agree that dominance and alpha animals are real." However, he agrees with Mech that in wolves, dominance is asserted in very subtle ways rather than in shows of violence, and is mainly used to decide order when distributing food to offspring. A far cry from so-called dog experts like Cesar Milan and their interpretation of alpha animals as bullies! Power issues cannot be dismissed out of hand, but nor should they be seen as a be-all and end-all. Research on wild packs of wolves has shown that cooperation is more important than asserting leadership, and alpha animals (parents) will choose to give food to pups rather than older offspring when food is scarce (Mech 1999) (Figure 3.3).

Figure 3.3 Pack of grey wolves in Yellowstone National Park. In this pack, cooperation is more important than maintaining a fixed hierarchy. Photograph by Doug Smith, National Park Service.

But the question raised by Cudworth goes beyond such pack hierarchy in the wild, and enters the minefield of modern, industrial farming. Although her question can seem irrelevant for archaeologists, that study past human–animal relations and do not have to concern themselves with the workings of so-called factory farming, I would argue that archaeology, too, should consider this question. Embedded in the notion of power is the power to take away rights of individuals, such as the right to choose, and the right to freedom. On farms, past and present, relations are managed by the farmers—and also by the domestic animals, both with their specific requirements to thrive and form a healthy flock, and by the animals' ability, by their agency, to choose whether to cooperate or resist.

Further, considerations of past farming is one of the areas in which archaeology has been biased by current narratives (Armstrong Oma 2013c). Modern farming has had an impact upon the epistemology of archaeology, stemming from the days of functionally-minded positivist processualism (e.g. Binford 1981; Higgs 1972; for critique, see Grant 2002). As discussed in chapters 1 and 2, it has a bias towards cost-efficiency and maximization strategies with a single-minded focus on economic gains. The rationale of the model rests upon a short-term perspective, underlying most farming strategies in the Western world (Alrøe *et al.* 2001; Dabbert *et al.* 2004; Lund 2002; Lund *et al.* 2004a, 2004b; Lund and Röcklinsberg 2001). Regarding husbandry practices, people in the past are often portrayed as being driven by maximization strategies similar to today's agricultural aims (e.g. Zimmermann 1999b). In archaeology, ambitions for the outcome of farming practices are frequently projected onto past societies. For example, very little consideration for animals is seen, and no affinity with animals is recognized. Animals are meat and calories (Albarella 2001; Grant 2002; Hamilakis 2001), and dominated by human power strategies.

From a theoretical perspective, one way of addressing and countering this bias is by approaching human and animal agency as a way of inter-species being in the world termed mutual becomings. This integrates factors such as situated knowledge, embodiment and the materiality of the body, and relationality. This idea is formulated and explored by Birke, Bryld and Lykke (2004). By using the feminist concept of performativity, as developed by Judith Butler (1993), they see animals as having agency by their ability to act upon the world. This creates an understanding of animals as a doing or a becoming, formed by social interactions. From this springs the idea of human–animal relationships as a mutual becoming, in which becoming cannot be separated from becoming-with.

As outlined above, a field of action—intra-action—forms a bridge between humans and animals, and allows them to develop together through a process, mutually constituting each other. In this field of action, humans and animals are engaged in mutual decision-making, a co-creation of behaviour that creates a choreography. In this fundamentally feminist understanding of the world separate groups of agents are not interesting, but rather focus is on the dynamics in-between them, and how this dynamics initiates changes. Although power strategies and hierarchic structures are acknowledged as present in the situation, they are not the heart of the matter.

Feminism and animals

Human–Animal Studies and feminism might seem like uneasy bedpartners, but scholars with feet in both camps have successfully drawn upon the theoretical frameworks of both. Examples are Lynda Birke, Donna Haraway and Ericka Cudworth, that have demonstrated that feminist theory and Human-Animal Studies have common interests and can fruitfully cross-fertilize. Lynda Birke (2002) has pointed out that women and animals have in common that they have both been oppressed, and there are movements to defend their rights; "Both feminism and causes for animals must share a concern with the way that the Other becomes subordinate" (Birke and Parisi 1999). Ecofeminism points at important experiential, theoretical and linguistic parallels between oppression and abuse of women and nature (Cudworth 2011, 24). Nature, in this respect, stands for wilderness and wildlife—hereunder animals—in the Western cultural tradition. Ecofeminism critiques the foundation for these parallels that, inspired by deep ecology (developed by the philosopher Arne Næss), is connected to the current threat from Western progressive resource mismanagement.

Regarding the archaeological study of animals, I find the common denominator between Human-Animal Studies and feminism to be the central precept of moving away from the uniform to the multidirectional. For feminism, this has meant shedding the idea of one type of agent as the norm (the white, middle-aged, middle class man), and for Human-Animal Studies it has meant to move away from speciesism. Speciesism is discrimination, a term used to denote a prejudice against non-humans based on physical differences that are given moral value, the assigning of different values or rights to beings on the basis of their species membership (Ryder 1971). This implies that animals are subjected to a bias in which they are perceived as less than humans, thus justifying human use (and abuse) of animals as we see fit.

Contrary to speciesism, feminism is useful to Human–Animal Studies because it has introduced a wide spectre of agents, and many feel that rather than adressing sexed agents as one or the other, man or female, we should think of sex as translated into a gendered process, where there is a scale of masculinity and femininity, that also includes age (see for example Gilchrist 1999).

In historical times in the Western world, women and animals arguably share a history of oppression (Birke 2002, 2010). Feminism grew out of a growing discontent with the suppression of women, and has, in certain periods of history, crystallized in issues such as the right to vote, taking steps towards assuring that the voices of women are heard. Today, feminists identify issues such as equality regarding childcare and wages. Similarly, animal advocates have given a voice to animals, and stood up for animal rights. The animal rights movement has become a powerful voice, as demonstrated by organizations such as Compassion in World Farming (CIWF), PETA and RSPCA. These organizations have targeted farming practices that are considered cruel to animals, and have for example promoted humane butchery practices, free-range or increased size of cages for laying hens, and tried to ban the use of lab animals in the cosmetics industry. Thus, women and animals have a common history of fighting for rights, and have their own "rights" movements (Birke *et al.* 2004, 176).

Interestingly, another common denominator is that both the women's rights and the animal rights movements have a shared history of deconstructing an androcentric bias. Arguably, for animals, the bias is essentially anthropocentric, but behind anthropocentrism is the normalized agent—the man. Throughout Western history, man has been the norm that all else has been measured up against. The (idealized) *man* embodies rational thought, knowledge and the ability to take the appropriate action. In the social sciences this led to the man becoming the defining subject—both the man as the measure of all things, and also the man as subject that defines all other beings as objects. Not only are animals objectified within this bias, women are, too. The idea of man as the action hero was also transferred to the animal kingdom. For example, in biology it led to biassed studies of primates, where male animals are seen as dominating aggressors. Women and animals were construed as objects for men to exploit in man's evolutionist search for power and progress. Both women and animals were subordinate "others," postulated to be closer to nature, the uncivilized, wild and mindless. A huge body of literature documents such a bias, and early feminist works within archaeology did an important job in deconstructing the idea of man as the sole subject (see for example Engelstad 2001; Gero and Conkey 1991; Gilchrist 1999; Stig Sørensen 2000). Likewise, feminist biologists focussed upon male bias in studies of animals. In archaeology, this eventually led to a worldview of greater complexity, with more than one human gender being studied. Feminism has in this way been imperative to introducing a variety of agents into the social sciences. From a Human-Animal Studies perspective, it is notable that among these agents can be found animals. However, it should be noted that feminism has not always been inclusive of animal agents and has not been at the forefront of recognising other species, and is not exempt from the prevailing speciesist bias (Birke

2012). Important to hold in mind, nonetheless, is the potential that feminist theory has when it comes to animal agency.

The notion of animals as agents has some interesting implications for feminist gender studies. Birke, Bryld and Lykke hold that "engagements with animals are also seminal to the gender process, since animals allow for a different perspective on the questions of agency and the importance of bodies; why bodies matter" (Birke *et al.* 2004, 176). In my own research, I have suggested that considering animals having the ability to act as significant Others (see also Cudworth 2011), a new type of agent is introduced in the gender process. Seen from this perspective, the engagements between individuals and animals—or animal materiality—reflect (gendered) identities (Armstrong Oma 2007, 260). Such processes, engagements and negotiations are all about human–animal relationships.

Feminism and Human–Animal Studies crossfertilize successfully when it comes to relationality. A useful concept in this regard is Donna Haraway's notion of companion species, and what happens "when species meet"—where she opens our eyes to other kinds of being as companion species, and the messy intersections that result from such meetings. According to her, animals are here to live with, not just to think with—and as companion species they do just that. She talks of "becoming with," and conceptualizes companion species as "cum panis," with bread, "messmates at table are companions" (2008, 17). Messy mates! This touches at the heart of my interest in the whole messiness of animals and their entanglements (gender being one of them)—I want to examine and understand living-with, becoming-with, and the social practice shared between humans and animals. Haraway succinctly spells this out with regard to dogs; "[d]ogs are not surrogates for theory; they are not here just to think with. They are here to live with" (2003, 5). In archaeology this could mean conceiving animals as producers rather than products and being active participants in the household (Armstrong Oma 2007, 2010, 2013b, 2013c; Årlin 1999).

The critical aspects of feminism and its critique of the androcentric worldview naturally lead to posthumanism, that challenges not only an androcentric, but a more encompassing anthropocentric world view. It is therefore to posthumanism that I now turn, a theoretical development that holds the potential to turn culturally established "truths" on their heads.

Posthumanism

The budding theoretical field referred to as posthumanism is almost easier to define by what it is not: humanism (Tsing 2012; Wolfe 2009). A common ground is the critique of the humanist project, and a desire to move away from anthropocentrism and ideas such as the culturally normative fantasy of human exceptionalism. Humanism is a child of the enlightenment ideas of Descartes and based upon the cultural historical inheritance of Christianity, in which humans are at the top of the hierarchy of species. Archaeologist Christina Fredengren summarizes posthumanism thus:

> as the questioning of anthropocentrism and the ontological divide between humans and nonhuman others with a focus on the interrelation between different species and the surroundings, arguing for a nature-culture continuum. By mapping a variety of embodied subject positions it raises issues about power, ethics and accountability from a standpoint where subjects have multiple belongings. (Fredengren 2013, 55)

Posthumanism situates itself as the driftwood of the smashed ship that is humanism, and it has mostly been defining itself, as Wolfe (2009) claims, by what it is not. One of the consequences of this is that there are many ways of understanding the rationale of large systems, such as societies, ecosystems, etc. Also, the shift from humanism to posthumanism demonstrates how such -isms are not written in stone. Thus, normative science and its rationale is not the only way of looking at the world (for example Midgley 1996). For example, an ontological hierarchy of species is not an objective truth, but rather a particular way of looking at the world. Another way of conceiving the world is a flat ontology (DeLanda 2005) regarding species, in which none hold supremacy beyond their biological abilities.

Although some academics that work within Human-Animal Studies are oriented towards posthumanism, this does not mean that Human-Animal Studies is a posthumanist field of research. However, fruitful synergies can take place. Certainly, the decentralizing of the human, which is such a central tenet to posthumanism, allows Human–Animal Studies the space to look at animals as themselves. Human-Animal Studies and posthumanism frequently meet some common fears. One concern that is often voiced in relation to posthumanism is founded in an irrational fear that the separation between humans and animals must be upheld lest animals somehow "take over," take control, and drive human society into utter chaos—much like the pigs in George Orwell's book *Animal Farm*. Another response is that giving animals more rights will detract rights from others, such as women, children, ethnic minorities and those with disabilities. Like the political philosophers Will Kymlicka and Susan Donaldson (2011), I do not think think that more rights to one kind of being will take away the rights that other groups have won. In fact, history demonstrates that the opposite is true, for example: the abolition of slavery didn't prevent women's right to vote nor child protection acts.

Posthumanism has grown from various critiques of the central tenets of humanism, namely setting the human species at the top of the hierarchy of creation. Some philosophers refer to this as speciesism (Berger 1971; Ryder 1971; Singer 1975), implying that animals are subjected to a bias in which they are perceived as less than humans, thus justifying that humans use (and abuse) animals as we see fit. My incentive here is not to argue against such a bias, or for animal rights in contemporary society, but rather to take wisdom from these philosophers in order to rethink the way archaeologists study animals in the past. Decentralizing the hierarchy of species—or thinking of it in terms of non-hierarchy—is relevant to archaeological forays into the pre-Christian past, in which species, kin and kind was construed according to different rules compared to the present Western world with its Cartesian mindset. Such a starting point can render new and unexpected insights.

Posthumanism holds a potential for openness and a moving away from anthropocentrism that allows for explorations of animals on a previously unforeseen level. Interesting for my project, which is the search for a means to address and identify human–animal relationships, is the insight that there is no inherent hierarchy amongst species where humans are at the top. Rather, posthumanism is interested in exploring the abilities of other species with an open mind. My initial response to posthumanism was a great interest in it, and I thought it might solve some of the ontological problems that I have struggled with, such as ways to approach animal agency, the question of animal consciousness and sentience, and an antidote to speciesism.

As an overarching framework, posthumanism holds great potential. It sets out to banish human supremacy, and tackles the Cartesian *cogito ergo sum* head on. This has led to a development of academic works that are truly profound and thought provoking. However, posthumanism has yet to fulfill its potential, because in practice, the theorizing has not led to much, and posthumanism now seems bent on a road that is free from moral concerns. Instead of embarking upon the path to wrestle with the big philosophical questions, entangled and interelated webs of technology and forms of life are delved into, and the messy hybrids born out of modern culture are investigated at micro-level. An example pertaining to webs of hybridity between humans and non-human animals are the use of pig's valves in human heart surgery. Posthumanism opens the door to consider machines, robots, cyborgs and animals as kinds of agents that have intrinsic value and are worth studying in themselves. Posthumanism goes into the realm of ethical questions of the fringes and borderlines—questions that have a huge bearing on future bio-politics. Such questions challenge the boundaries of not only species but also the hybridity between humans, animals, environment and machines. Examples are issues such as cloning, stem cell research, artificial intelligence, and cyborgs. Popular media images are for example the so-called earmouse, or Vacanti mouse, with an ear grown onto its back.

Feminists such as Donna Haraway (2003, 2004, 2008), Erica Fudge (2008) and Erika Cudworth (2011) are cautious of posthumanism and shy away from identifying with it. Haraway has expressed a concern with posthumanism because she fears that it is veering towards a utopian transhumanism, in which animal parts become a tool for unaturally extending human life. And, as stated by Cudworth (2011, 8), "transhumanism is [...] not interested in animals very much at all, be they human or otherwise." For this reason Haraway has abandoned the cyborg in favour of "companion species." Similarly, a posthumanism that turns towards transhumanism is not useful as a theoretical framework for the kind of archaeological investigation that turns towards human–animal relationality. Rather than opening up new approaches to study animals and humans in an interrelated web, animals become props for human technological antics. Rather than investigating animal consciousness, animals are celebrated for their partability. Hardcore biology provides DNA data that tells us about a species' potential and aptitude but is mainly used to test the capabilities of animal parts to human medicine. Thus, human supremacy was never knocked off the throne.

Rethinking how archaeologists study animals in the past: Towards a feminist prehumanism of human–animal relationality

A posthumanist archaeology is in some ways an oxymoron, since archaeology for the most part deals with societies that were, by definition, prehumanist. Prehumanism is a more appropriate term to theoretically deconstruct humanist narratives in archaeology. After all, the subjects of investigation mostly lived long before Descartes and his contemporaries, and we can safely assume that the humanist ideal of the enlightenment didn't enter into their sphere of reasoning. In his book "What is posthumanism?" Cary Wolfe tackles this question. He suggests that it comes both before and after humanism—before "that historically specific thing called 'the human' that Foucault's archaeology excavates." From an archaeological perspective, what is needed is to go beyond humanism and scientifically challenge its inherent ego/anthropocentrism. This would be a true landslide in the history of science (similar to Darwin and Copernicus), and is one that is begun in other disciplines by Marc Bekoff, Donna Haraway, Erika Cudworth, and Lynda Birke, to mention some.

I have found feminism to be a very useful tool to approach such understandings of inter-species relationships. Maintaining a feminist starting point when studying human–animal interactions in the past merges into an approach that could be termed feminist prehumanism: within such a framework, animals and humans are seen as together engaging in a dialectic relationship where both affect each other—and animals have the capacity to partake in a mutual creation of life-worlds. Within such a framework, the human–animal relationship is a process between individual humans and animals. The act of relating is a course of encounters where different agents get together and are situated at an interface. Relating is to interact, and requires reciprocity, exchange and proximity. Thus, it is an act where one is embedded in the world together with others; humans and non-humans share the world and relate to each other in it. This mutual engagement is born from the co-habiting situation of sharing the world, as humans and animals alike dwell in the world. This is how relationships happen.

From feminism, the valuable lesson learnt is that many kinds of being can be agents. Understanding how relationships form means that all participants are taken seriously (Birke and Hockenhull 2012a, 16). But herein lies the crux: the legacy from the Western cultural history, and the development of humanism in particular (as described above) have put science in a position in which researchers have been conditioned to study the human species in a way that takes itself—and only itself—seriously. However, I wish to take other participants seriously, too. By doing so, some new questions concerning animals and agency can shed light on processes that hold great time-depth. We can ask what kind of impact animals have had on past societies, beyond as a source of meat and calories. And further, can a consideration of animals as themselves, and the way they live and interact with humans within shared worlds, help with understanding the human cultures that lived or live with them? In the following chapters these ideas are tested as the case studies are discussed both considering their archaeological context and how human–animal relationality emerges when considering a Human–Animal Studies perspective.

The ontological turn: Exploring the ontological status of animal agents

Taking other-than-human animals seriously as participants in forming relationships (e.g. Birke and Hockenhull 2012b) entails not collapsing them within a single frame of reference—i.e. as The Animals. Although animal is useful as a broad category, to truly understand and appreciate their different roles, I suggest that their ontological status within historically specific societies should be studied.

Ontology is the study of, and inquiry into, that which is. By this I mean that it grapples with the very foundational building blocks of the world—and moreover, how these building blocks are perceived by different cultures as the natural order of the world. Ontology is, then, not by its own nature fixed and stable, but it appears so by the way it is situated in time-space, within its own historical-cultural context. An example is the Cartesian worldview that has been put under scrutiny in this and the previous chapter, in which humans were considered as beings with souls whereas animals were seen to be like machines.

Recently, there has been a return to inquiry into ontology in anthropology, and to a minor degree in archaeology—frequently referred to as the ontological turn. Proponents of this school of thought have emerged from the outside of the Anglo-American anthropology, foremost by way of Eduardo Vivieiros de Castro, Eduardo Kohn and Bruno Latour. For example, in his groundbreaking book "How forests think," Kohn (2013) explores ontology as an interconnected web of emergent meanings. It is the web of meanings that give ontological status to beings. This is often termed a relational ontology, meaning that ontological status springs from the sum of threads in the web—what Ingold refers to as meshwork. A bounce on the web travels along the threads and affects various entities.

Thus, a core idea situated in the midst of the ontological turn is that the nature of the world (ontology) is constructed in the same way that knowledge (epistemology) is contructed. In a discussion article on the ontological turn, Severin Fowles (Alberti *et al.* 2011, 898) argues that ontology springs from an understanding of origins and that ontology is anchored in narrative; "The world is as it has come to be." Instrumental to this development has been Eduardo Viveiros de Castro and his work on perspectivism amongst Amerindians (e.g. 1998, 2004). In his case studies, he is able to demonstrate that humans and animals are interchangeable, and that humans can take on the perspective of animals, i.e. be in the world through the body of an animal. The narrative of origins of these people establishes that

> many animal species, as well as sundry other types of nonhuman beings, are supposed to have a spiritual component that qualifies them as "people." Such a notion is often associated with the idea that the manifest bodily form is an envelope (a "clothing") that conceals an internal humanoid form, usually visible to the eyes of only the particular species and of "transspecific beings" such as shamans. (Viveiros de Castro 2004, 465)

Thus, Amerindians perceive animals as posessing an inner, human form that is hidden by their bestial bodily form. In comparison with the Cartesian worldview, the nature of being is turned on its head: humans exist in animal form, and animals exist in human form—and the diversity of animal bodies has evolved out of humanness. I argue that his finding opens up the field of ontological inquiry and makes possible a range of understandings of agency, personhood and ontological status in all societies, also those of the past. I follow Benjamin Alberti's statement (2011, 900), "I conceive of ontological inquiry as a means to insert a difference [...] in the present and in our accounts of pasts."

I will not probe further into the theoretical arguments that spring from the ontological turn, as launched in anthropology. Suffice to say that ontology—the study of existence itself, the nature of being, becoming, reality—is at the core of my investigation of various animals and their agency. In particular, I explore what kind of ontological status various kinds of beings have/had, and how the agency of these beings lent them a particular ontological status. By ontological status I mean how the different being's place in the web of life was understood and formed by the emergence of particular—acted out as well as materialized—relationships. Returning to the idea of meshwork (Ingold 2011), relating—and shaping ontological statuses—emerges out of the lines that connect the web. The way life-worlds are and were co-shaped between entities lent those entities their ontological status, by mutually becoming in a back-and-forth interchange of action, situatedness and conviviality.

In the following chapter I delve in-depth into the case study, to examine in depth how a study of houses, households and the environment in the Bronze Age can "insert a difference" (Alberti 2011, 900) into how we make sense of this segment of the past and the beings that populated it.

— 4 —

Three-aisled Houses in Early Bronze Age Rogaland: Who were the Household Members?

Changed way of building, changed way of living

In the Early Bronze Age, a monumental change in the planning and building of houses happened, leading to the transition from two-aisled to three-aisled longhouses. Architectonically, the construction of the longhouse changed from three rows of post-holes with one central roof-bearing row, into four rows with two rows of central roof-bearing posts. In this transition, the three-aisled longhouse was created (see Figure 4.1). This architectural change is by some archaeologists believed to be associated with a change in how animals were kept; they were moved into the three-aisled house and lived in the same house as humans (Årlin 1999; Lagerås and Regnell 1999; Rasmussen and Adamsen 1993b, 138; Rasmussen 1999, 281; Tesch 1992, 290). The transition is normally dated to Bronze Age period II (Montelius), see Table 4.1 below. Factors that caused this change are not known, nor properly accounted for in the archaeological discourse. Chief questions are: How did humans and animals live together in Early Bronze Age households? Were humans and non-humans both members of the household? If so, how did this manifest? What was the ontological status of these beings?

In this chapter I examine the background to the case studies as well as the case studies themselves in detail, in order to examine the basis of species membership in the household. I look specifically at sites in Rogaland, since it has a large compilation of studies, with datasets that have been thoroughly examined and are well presented. The region of Jæren and adjacent areas dominate the data, since development has been most intense there. I look specifically at the development of a change that occurred in ways of shaping the house, from two-aisled to three-aisled longhouses over a delimited period of time, Early Bronze Age period I–III. Studying a development over time makes differences evident.

A closer look at the wider context of the transition from two-aisled to three-aisled houses is required in order to provide background to the case studies. In the following, I will present the postulated architectural change that took place, both regarding its place in the archaeological discourse and in terms of changes in husbandry practice and land use. Subsequently, case studies from Early Bronze Age Rogaland are presented

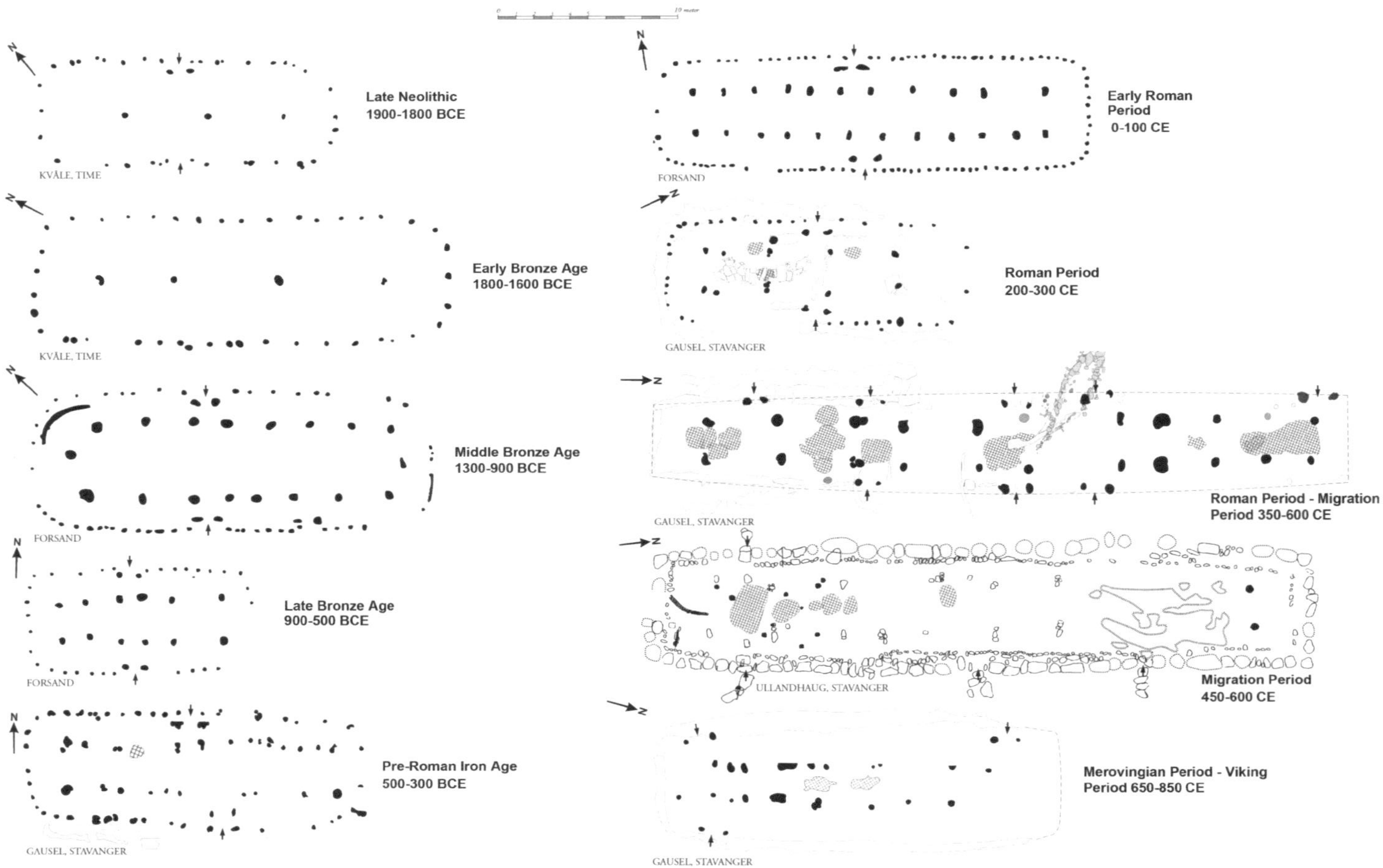

Figure 4.1 The development of longhouses in Rogaland from the late Neolithic until the Viking period, a period that spans 3000 years. Note the transition from two-aisled to three-aisled that happened early, at the 1800-1600 / 1600-1400 transition. Illustration by Ragnar L. Børsheim / Museum of Archaeology, University of Stavanger.

Table 4.1 Correlation of Montelius' periodic system with radiocarbon dates. Note that transitions are not always clear cut in terms of dating.

Period	EBA I	EBA II	EBA III	LBA IV	LBA V	LBA VI
Time (BCE)	1700–1500	1500–1300	1300–1100	1100–900	900–700	700–500

and the evidence is discussed in tandem with developing the argument pertaining to an increase in sheep-rearing. Social and practice-oriented implications, as well as environmental and economical inferences, will be outlined. Thereafter, I shall use my theoretical background as a human–animal scholar to interpret the implications of these changes in terms of human–animal relationships.

A common assumption is that a result of the architectural change was indoor stalling of domestic animals in one part of the house. Some archaeologists (Årlin 1999; Armstrong Oma 2007, 2010, 2012, 2013b, 2013c, 2016a, 2016c; Lagerås and Regnell 1999; Rasmussen and Adamsen 1993a; Rasmussen 1999, 281; Tesch 1992, 290) have previously suggested that the change happened due to a change in perception of domestic animals, leading to them becoming household members and embedded in the life-space of humans. Thus, a more intimate human–animal relationship developed. This is aptly demonstrated by the remains of a house that burnt some centuries later in Nørre Tranders on Jutland, in which the bones of animals were unearthed *in situ* in the part of a longhouse interpreted as the byre (Nielsen 2002). This house was dated to the Pre-Roman Iron Age, and is one of several from this period around the Limfjord area on Jutland that were burnt, where the remains from humans and animals were discovered in the position they perished in (Kveiborg 2009).

Another analogous site is found at Must Farm[1] in the fenlands near Cambridge where recent excavations have unearthed a "Pompeii situation" where a settlement from the Late Bronze Age (1000–800 BCE) with several stilted houses collapsed suddenly into the river due to a fire, leaving humans, animals and material objects in situ preserved by a layer of silt (Marchini 2016). In their online dig diaries, the excavation team, led by Mark Knight, describe preliminary results of finds within the houses: "The animal remains found underneath the footprints of the stilted houses are associated with the buildings' destruction. They are found alongside other burnt material and were probably inside the structures when they collapsed into the river beneath the homes. Overwhelmingly, these animal bones are domestic and dryland in character, dominated by semi-articulated young sheep and calf remains."[2] The semi-articulated sheep and calf bones could represent carcasses that were cured by hanging from the rafters, but considering that sheep and cow droppings were found both inside and outside the houses, the remains probably represent living animals that were inside the house.[3] Lamb bones were found in the western part of the houses, together with some socketed axes but nothing else, whereas in the eastern part large amounts of household implements were

1. http://www.mustfarm.com/bronze-age-settlement/
2. http://www.mustfarm.com/progress/site-diary-35-the-must-farm-menu/ accessed December 13, 2017.
3. Personal communication Mark Knight, November 10, 2017.

scattered (Marchini 2016). This points to a division of internal space similar to that seen in longhouses. The excavators further note that other excavated roundhouses also frequently have sheep remains inside or at the entrance to the houses, "in fact, it is very rare to find roundhouses of this period at dryland settlements that do not have young sheep remains inside."[4] Outside of the houses, the faunal remains mainly consist of wild animals and the bones are butchered and dispersed. Remarkably, many fragments of textiles were preserved and these were made by plant fibres rather than by wool.[5]

Although I firmly believe—and evidence from later periods from the Limfjord region confirms—that houses shared between humans and animals became the outcome of the changes in ways of living, it does not account for what instigated the process of granting the animals access to the house, and ultimately ontological status as household members. Such changes are most often complex conundrums made up of a number of factors. I suggest that in Early Bronze Age Rogaland, sheep were at the heart of such a conundrum, and in this chapter I will present archaeological, environmental and pictorial datasets to examine this line of argument. Why sheep appears to have taken the centre stage amongst the domestic animals is unknown. However, I argue that the innovation of wool-based textile production led to a favouring of sheep, and that rearing sheep for this purpose led to a profound change in ways of living and being with other species—most notably sheep and dogs. Giving sheep primacy in the Early Bronze Age is contrary to the common assumption that cattle were the most important and most abundant animals (e.g. Rasmussen 1999; Holst and Rasmussen 2013). Although the dominance of cattle is probably accurate for Denmark, I present data that suggests otherwise for Rogaland.

Three-aisled longhouses: The knitty-gritty

Two-aisled longhouses are interpreted for human habitation only, based upon the lack of internal divisions, which signifies that the whole house was used for human habitation (Figure 4.2; Ethelberg *et al.* 2000; Tesch 1993). In her analysis of charred grains from the Neolithic and the Bronze Age in Norway, botanist Eli-Christine Soltvedt (2017, In prep) notes that the two-aisled houses have large amounts of charred grains in the postholes, whereas the early three-aisled houses have almost no charred cereal grains in the postholes. Obviously, patterns of storage and deposition play a role concerning the amount of charred remains in postholes, but considering the amount of grains in some of the postholes of the two-aisled houses, I suggest that it is likely that the grains were intentionally deposited. Further, the fact that the amount of grains is low in the postholes of three-aisled houses suggests to me that grains were no longer as important and probably also that there was less of it around. Thus, the change in amount of grains from two-aisled to three-aisled houses can be understood as reflecting a change in economic strategy—from a focus on growing grains to a focus on pastoralism.

The development of architectonic articulations is rather abrupt, with one type of house—the two-aisled—supplanted by the three-aisled longhouse. The latter was rectangular and often large, varying between 50–20m long and 12–7m wide (Harding 2000,

4. http://www.mustfarm.com/progress/site-diary-35-the-must-farm-menu/
5. http://www.mustfarm.com/bronze-age-settlement/about/

Figure 4.2 A reconstruction of a two-aisled longhouse. Even though there would be ample space for animals inside the house, there is no evidence to suggest that animals did live inside these houses. Illustration by Ragnar L. Børsheim / Arkikon.

45–48). The transition from two-aisled longhouses with exclusive human occupation to a shared living-space where humans and animals occupy different parts of the three-aisled longhouse occurred in the period between 1800–1500 BCE in northern Europe (Rasmussen 1999). Traces of byres and internal divisions appear c. 1500 BCE (Fokkens 1999, 36). These houses are probably designed to stall animals indoors (Ethelberg *et al.* 2000, 203) which is the main reason that the transfer from the two-aisled to the three-aisled longhouse is associated with moving domestic animals into the house.

In the Neolithic (c. 4500–3800 BCE) Linear-Band Keramik (LBK) houses in central Europe, cattle might have lived in one part of the house (Marciniak 2005, 14; Startin 1978). However, there is both a big time-gap and a big geographical gap between Central Europe in the Neolithic and Northern Europe in the Early Bronze Age. It is safe to assume that no cultural connection existed, and thus the three-aisled longhouse was, in Northern Europe, an original development.

Three-aisled longhouses became common in southern Scandinavia, Northern Europe, the Low Countries, at the Alpine lake sites and all over the Urnfield "world" (Harding 2000, 38). The break in the conservative building tradition of two-aisled longhouses have been suggested to represent "radical or gradual changes in society" presumably of economic, social and political character, according to Tesch (1993, 145). This corresponds well with the changes in the depositions of charred grains in postholes, a prominent feature in two-aisled longhouses that ends abruptly with the three-aisled longhouses (Soltvedt 2017) In addition, following Rasmussen (1999) and Årlin (1999) I argue that this shift was, at heart, a change in the ways humans and animals perceived one another, related to each other and shared their world.

Early Bronze Age farms on Jutland

Bronze Age Jæren is often compared with sites and cultural traits found in the contemporary Danish territories, particularly Jutland, which would have been accessible across the sea. A network of exchange has been postulated, based both on great similarities in cultural expressions and on the rock art depicting ships, believed to represent real journeys (Kvalø 2002).

The Early Bronze Age farms in Denmark are often used as a default when considering settlement structure and house construction on Jæren, and comparative references to the Danish Bronze Age are frequently sought (Fyllingen 2015a, 2015b; Fyllingen and Armstrong 2012; Løken 1989; Soltvedt *et al.* 2007). Tacit assumptions that people in southern Scandinavia led the way often underlie older research. There is substantial evidence to demonstrate that the houses were indeed, mostly, constructed in the same way. But were their uses exactly the same?

Bronze Age farms with three-aisled longhouses have been extensively investigated in southern Scandinavia, particularly on Jutland. Byres are frequently identified based on the presence of clearly defined stalls in one part of the house. Not all excavated longhouses have physical traces of byres. The houses show great differences, and only a few examples have clearly defined stalls (Rasmussen 1999). Where byres are found, they are introduced as a visible and separate part. All traces of activities in these houses seem to respect the partitions of the rooms; this does not only hold for the internal divisions across the houses, but also the aisles partitioned by the presence of the roof-bearing posts (Rasmussen 1999, 283). The house types that most likely contain byres are long and narrow houses with partition walls where animals probably lived in the part interpreted as the byre (see Figure 4.3). Wider houses with partition walls indicate that larger groups of people lived under one roof. The trend clearly shows that houses with internal partitions where no outhouses are present contained byres for stalling animals, whereas smaller houses with outhouses show an arrangement where humans lived in the house and animals lived in the outhouses (Rasmussen 1999, 285). Thus, animals were, to a certain extent, brought into the house in the Early Bronze Age, but other buildings and structures were also a part of animal husbandry management.

However, this argument is based upon the frequently tacit (but sometimes vocalized) assumption that the domestic animals that lived in the house were cattle, located in stalls along the long walls (e.g. Rasmussen 1999; Horve 2009; Holst and Rasmussen 2013). The houses with clearly defined stalls do indeed demonstrate that larger domestic animals like cattle and horses were stalled indoors in some houses. Horses very rarely appear in the faunal remains (Nyegaard 1996), therefore the majority of the stalls were probably designed with cattle in mind. But, there are other ways of keeping cattle inside human living-spaces, such as free-range, tethering or penning. Also, stalls do not account for how smaller domestic animals, sheep, goats and pigs were kept inside. Penning them obviously makes more sense than stalling them. The distribution of bones and architectural features within the burnt house from Nørre Tranders and other, similar houses from the same period and region demonstrates that sheep and pigs were kept inside the byre but not in stalls (Kveiborg 2009; Nielsen 2002). Thus, I argue that houses with stalls are but one way of accommodating animals indoors, and like the house from

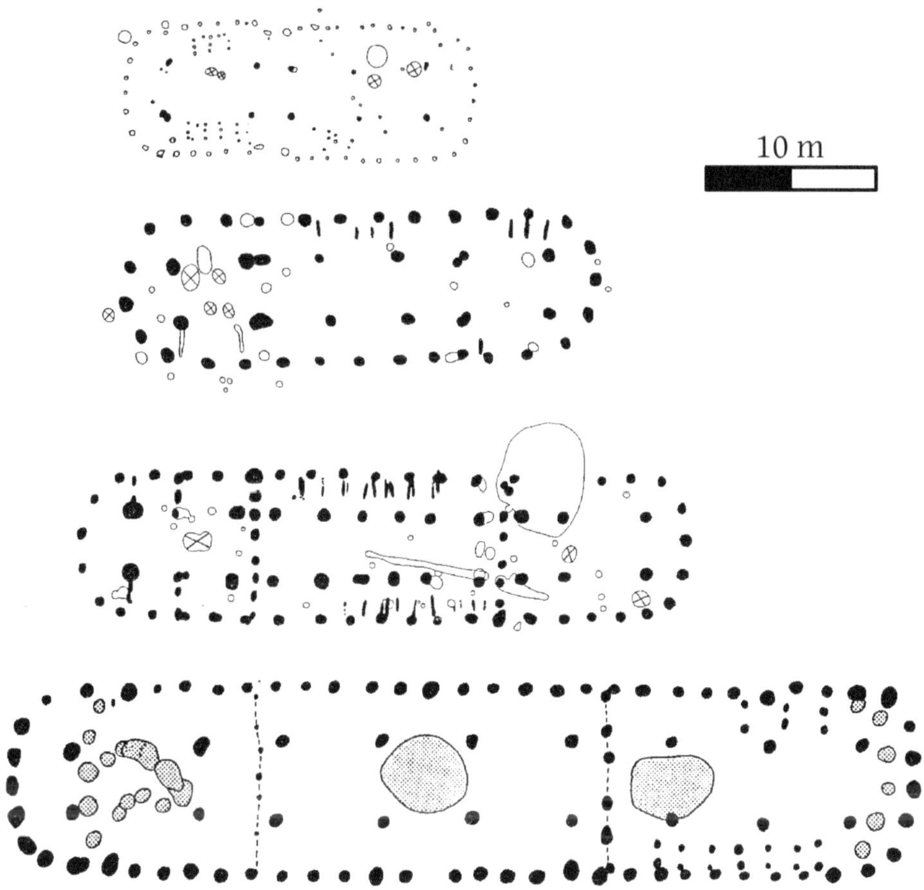

Figure 4.3 House plans from Jutland with possible byres in one section of the house. Note traces of stalls—smaller posts or remains of planks between the trestles and the wall posts. (Bech 1997; Ethelberg 1995; Mikkelsen 1997; Rasmussen and Adamsen 1993).

Nørre Tranders demonstrates, the lack of stalls does not necessarily mean that animals were not housed in three-aisled longhouses.

Three-aisled longhouses: The research discourse of indoor stalling

During the last twenty years, stronger emphasis has been put upon social processes in settlement patterns of the Bronze Age among archaeologists (Armstrong Oma 2007, 2010, 2012, 2013a; Arthursson 2009; Barker 1999; Fokkens 1998; Grön 2004; Olausson 1999; Rasmussen 1999; Roymans 1999; Streiffert 2001; Ullén 1996, 1997; Zimmermann 1999a). One of these is the innovation of sharing living space with animals and the resulting new aspects of the human–animal relationship. Usually, the transition is explained in terms of ways to facilitate everyday life for Bronze Age farmers. Archaeologist Harry Fokkens summarizes the explanations in the following three main functional arguments for stalling animals in houses (1999, 36):

- climatic reasons—protecting the livestock from the cold winters (Behre 1998, 94);
- a means of protecting cattle against raids (Harsema 1993, 106; Roymans 1999);
- enabling the collection of manure (Fokkens 1991, 1999; Gaillard *et al.* 1994, 60; IJzereef 1981; Karlenby 1994, 31).

To these, it is common to add two others:

- having animals in the house helps to heat it (but see Zimmerman 1999b).
- keeping animals indoors protects them from predators.

None of these arguments explain why it was necessary to bring the livestock into human living spaces, as these benefits could also have been achieved through other means, or are simply unnecessary. For example, Zimmerman (1999b) says that heat from animals does not reach beyond the byre part of the longhouse and has no effect on the human living section. On the contrary, the presence of animals creates a more humid and unhealthy indoor environment that leads the wooden posts and planks to rot quicker. The sturdy, small, quick domestic animals of the Early Bronze Age would be superiorly equipped to fend for themselves both against the weather—particularly taking into account that the Early Bronze Age was a very mild period, and against predators. As for collecting manure, it seems counter-intuitive to collect it inside and carrying it out rather than let the animals drop it where it is needed—in the pasture all year round and in fields where crops were grown during the fallow parts of the season.

Several authors (Årlin 1999; Fokkens 1999; Rasmussen 1999; Roymans 1999; Zimmerman 1999a, 1999b) argue that other factors were crucial in choosing to share the house with animals. Rasmussen (1999, 287) suggests that a "human to animal relationship" developed in this period, and advocates a point of view in which taking animals into the house was an expression of relations of affiliation or ownership of the livestock. In southern Scandinavia, very small faunal assemblages have been recovered from house contexts in this period (Nyegaard 1996); consequently, the evidence comes mainly from the architectural features of the longhouse. The visible internal divisions in the houses are important to understand the manifestation of the physical spaces created to accommodate the novel "human to animal relationship," as it suggests that animals were not perceived as humans. The physical separation of animals into a sphere apart, but still in the house, nonetheless signals that the animals were perceived as a valuable part of the household, and I suggest that as a consequence of being taken into the house, they were given status as household members. In terms of the ownership of animals it was defined by the building that housed them, and the building was defined by the humans living in it. Further, Årlin (1999) argues that, by living inside houses, domestic animals are perceived as categorically pure, following Douglas' (1966) structural separation between purity and danger as a cultural construction. Implicitly, animal products are perceived as pure, even dung, which gives animals the role of producers and, implicitly, as agents rather than as food and territory (Årlin 1999, 300).

Although this discourse refers to animals as a generic category, and sometimes to the slightly more specific, but derogative term, livestock, what is meant by animal is frequently cattle (Barker 1999; Fokkens 1999; Rasmussen 1999; Roymans 1999; Zimmermann 1999a, 1999b). For example, when considering the development of the byre, both Zimmerman (1999, 133) and Waterbolk (1975) discuss width of stalls and cattle size. Cattle is also clearly what Rasmussen (1999) has in mind when she suggests that a management strategy of animals is tied to agricultural activities, not only dung collection. The agricultural strategy of Early and Middle Bronze Age in southern Scandinavia seems to have been founded on an extensive cultivation, where new ground was constantly laid under the plough, requiring draught animals (e.g. Kristiansen 1993). The symbolic presentation of the relationship between humans and cattle on rock art always shows ploughing, but rarely pastoral activities such as herding (but see Holst and Rasmussens (2013) argument in favour of extensive cattle herding). Therefore, economic and social aspects joined together in creating a new way of living for humans and cattle in early and middle Bronze Age in southern Scandinavia (Rasmussen 1999, 288).

Considering domestic animals as household members suggests that husbandry was a very important task in the lives of Early Bronze Age farmers (Grön 2004, 370; Rasmussen 1999, 286). Rasmussen (1999, 287) suggests that the symbolic representation of a "human to animal relationship" is reflected in graves where the dead are laid on, or wrapped in, cowhide, and in animal offerings at settlement sites. Further, the new spatial organization of longhouses, practices surrounding burials and rock art demonstrates how pervasive livestock management had become in people's everyday life. Following Rasmussen's line of argument, I have previously argued that a particular world-view is reflected in the relationship between the people and their animals, and this world-view created the longhouse with byre, as we know it (Armstrong Oma 2013b, 2013c, 2016c). In the following I will give a background to, and subsequently present and discuss the archaeological data from settlements from the transition from two-aisled to three-aisled longhouses in Rogaland. This is done in order to explore these arguments on a case study that differs from the Danish and Southern Swedish material that the ideas presented above are founded upon. This will demonstrate that there may be a specific connection between the development of the byre and sheep management practices that might have been particular to Rogaland.

Bronze Age house typology in relation to Rogaland houses

The development of longhouses is surprisingly consistent throughout Early Bronze Age Northern Europe, and the typologies that have been created have a wide geographical applicability. Archaeologist Sten Tesch was the first to create a typology of three-aisled Bronze Age houses, based on a large body of material from Southern Scandinavia. His subdivision of Scandinavian Bronze Age longhouses from Scania includes 5 categories (Tesch 1993, 162):

- Long-houses: long buildings with no clear division into two sections.
- Long-houses: medium sized with a clear division into two sections.
- Long-houses: medium sized with three or four roof-bearing trestles[6] and no clear division into two sections.

6. The use of the word trestle is used for the Norwegian term *grindverk* which does not have a direct English translation, but means two upright posts joined by a horizontal timber forming a frame.

- Small houses with two roof-bearing trestles.
- Sunken floor huts.

The types described by Tesch (1993, 162) are roughly descriptive for all of southern Scandinavia, and are used on a regular basis in works that deal with Bronze Age houses (Borna-Ahlkvist 2002; Borna-Ahlkvist *et al.* 1998; Grön 2004; Horve 2009; Streiffert 2001; Tesch 1993). However, it is worth noting that the typology is based upon houses from the Köpinge area, and comprise of houses dating to Early Bronze Age period II–III and the Late Bronze Age. Houses from Bronze Age period I are not considered.

Archaeologist Inger Horve (2009) has created a typology for Bronze Age houses from Rogaland comprising of three basic house types (types 1–3 below). In addition, I add one more type (type 4), which has recently come to light.

- Type 1: Two-aisled longhouses, with Bronze Age dates.
- Type 2: Large and wide three-aisled longhouses, stretching beyond 7 m wide, with a span greater than 3 m between the trestles.
- Type 3: Smaller, narrow three-aisled longhouses, where the posts of each trestle are less than 3 m apart, and narrower than 7 m.
- Type 4: Hybrid longhouses where one part is two-aisled and the other part is three-aisled.

I have chosen to follow Horve's critera for house types 2 and 3. Although these types do not differ all that much regarding length and total area, the distance between trestles and the width of the house increases or decreases the inner space, and the feel of spaciousness (see Appendix).

In addition to these types, there are various other settlement types dated to the Bronze Age, smaller 4 or 6 post houses with or without hearths, as well as open sites and caves and rock shelters (e.g. Horve 2009; Prescott 1991a, 1995, 1999). Smaller 4 or 6 post buildings are normally not given priority when dating features from a site, therefore little is known of their timeframe (personal comment Hilde Fyllingen). Tesch's small houses with two roof-bearing posts are also found in Rogaland (most notably at Myklebust, see Dahl 2014). However, these are not longhouses and their use is unknown. They are therefore not considered in the following. Figure 4.4 maps the sites discussed in the text in relation to each other, and includes both house types 1, 2, 3 and 4.

Type 1 houses are found throughout the Late Neolithic, but houses dated to Bronze Age period I are considered here, since they are close to the transition to three-aisled longhouses. This house type does not continue in Rogaland after the three-aisled house was introduced. Altogether six houses of type 1 are considered here. Bjorhaug is included due to its excellent preservation, Kleppestemmen, Tjora and Kvåle because they are directly linked to subsequent habitation in three-aisled houses, and Talgje due to the unusual deposition of loom weights in a posthole. The houses vary greatly in size, length varies from 10 to 23 metres, with an average length of 18 metres, size varies from 48 square metres to 166 square metres, with an average of 113 square metres. Worth noting is that type 1 houses occurs infrequently all through the Bronze Age in other parts of Scandinavia, although none are found later than period I in Rogaland.

Some archaeologists propose that internal divisions in two-aisled longhouses are early indications of dividing the houses into a human living area and a byre-part (Fokkens 1991; Olausson 1999). However, it is suggested that this cannot be the case for the Rogaland, and Norwegian, houses (Sæther 2007, 15). Conversely, I argue that there might be pointers towards internal divisions also in two-aisled longhouses, for example house 1 at Kleppestemmen and house 3 at Bjorhaug, but the material is too flimsy to draw any firm conclusions. There are certainly no other pointers towards an internal separation into a byre and a human living section.

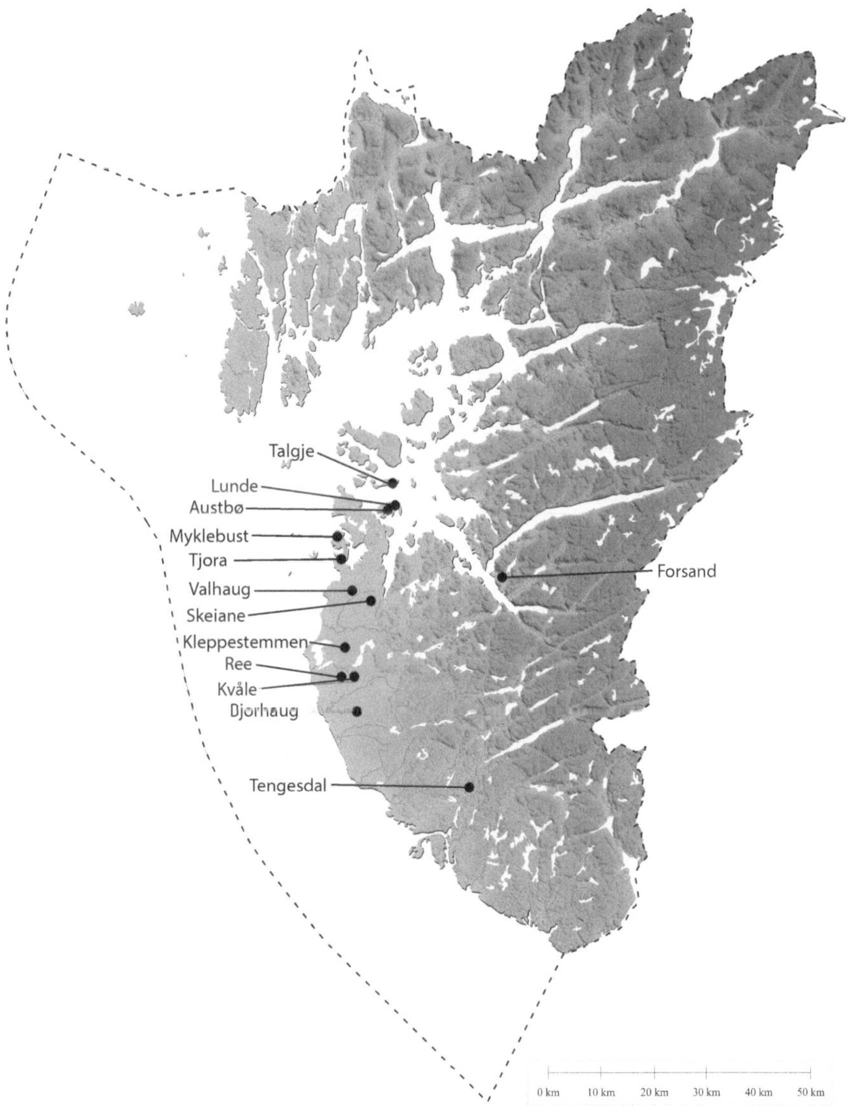

Figure 4.4 Map of Rogaland with Early Bronze Age shoreline and the spatial distribution of the sites discussed in the text. Map: Theo Bell Gil (graphics) and Lisbeth Prøsch-Danielsen (plotting of shoreline) / Museum of Archaeology, University of Stavanger.

Type 2 houses are found throughout the Bronze Age, dated to period II–V (Horve 2009, 30–31). Those dated to period II–III are considered here, altogether nine houses. These are: house II Kleppestemmen, houses I and II from Austbø lok 20, houses XXXIIB, LIX and XC from Forsand, house I from Kleppevarden, Lunde house 1 and house I from Skeiane. Compared to Horve's data set, only one house has not been included here due to its later date, namely House LXXV from Forsand. These types are often termed halls, due to their great width and large proportions. The average length is 20 m, and average size as large as 154 square m. However, some of the houses that are included here are wide, but not all that long. There is thus a certain variation within this group.

Type 3 houses have previously been dated primarily to Late Bronze Age period IV–VI (Horve 2009, 31), however my summary demonstrates that this type is also frequently found in the Early Bronze Age. Altogether seven houses of this type are considered here, house III from Kvåle, houses XLV and CCXII from Forsand, Myklebust houses VIII and XV, Tjora house 4, Tengesdal house 3 and house I from Skeiane. In comparison, altogether 13 houses of this type are found from later periods, nine from Forsand, two from Håbakken and two from Gausel. Average length of the Early Bronze Age type 3 house is 17 m, and average size is 101 square m.

Type 4 houses, hybrid houses, are dated to Early Bronze Age I (Fyllingen 2015b; Fyllingen and Armstrong 2012). So far, only two are found in Rogaland, both at Tjora, although one other from Tanangerveien is currently in the process of being analyzed (personal comment from Krister Scheie Eilertsen). They are dated to Early Bronze Age I, and are fairly small, average length is 13 m and average size is 73 square m. The type is also known from Denmark (Bech and Olsen 2013).

Considering the question of indoor stalling, houses of types 2 and 3 are most interesting. Can traces of byres be found in these houses, and if so, what kind of byres were they, and for which kinds of animals? Further—what of the hybrid houses—are they very early houses with byres? The sites and particularities of the houses are all described in detail below. The houses that make up my data are all listed in the Appendix. Six of those included here are two-aisled houses, but the majority are three-aisled houses. Of the latter, nine are of type 2, which according to Horve (2009, 30) are large and wide three-aisled longhouses, stretching beyond 7 m wide, with a span greater than 3 m between the trestles. Seven houses are of type 3, most of them considerably smaller and narrower than type 2. However, both type 2 and 3 are sufficiently large to contain both a human life-space and a byre for animals.

In the following, I will present and discuss most of the houses as well as their settlement context, except only Kleppevarden House 1, of which there is very little information. All figures are from the original documentation, some are redrawn, and consequently vary somewhat in quality and information. Environmental data and faunal remains will be described and discussed following a summary discussion of the houses.

Talgje

Talgje is an island in the Boknafjord close to the Stavanger peninsula and the Jæren region. The two-aisled longhouse is dated to a long stretch of the Early Bronze Age, spanning periods I and II. It is considered here due to its relatively late date considering

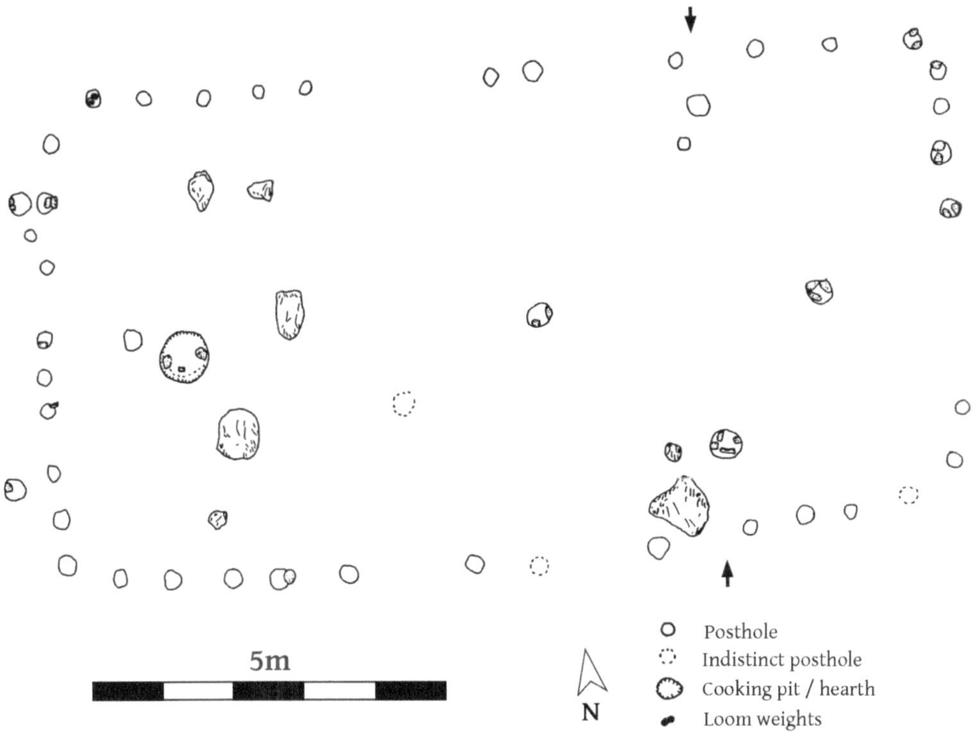

Figure 4.5 The house from Talgje. Loom weights were found in a posthole in the northeastern corner. Drawing by Olle Hemdorff (1993a, 24).

its construction and most importantly the find of loom weights inside the house. This is the earliest remnant of textile production in Rogaland, and is a strong pointer to the manufacturing of wool textiles.

The house is unusually shaped to be a two-aisled longhouse, as it is relatively short and broad (see Figure 4.5). Four clay loom-weights were stacked in the wall post of the very northeastern corner, filling the whole posthole. The excavator finds it inconceivable that a post was put on top of these loom-weights, there was simply no room and they would have been crushed. Instead, the loom-weights are interpreted as a sacrifice to mark the end of the life-cycle of the house (Hemdorff 1993b, 8; 1993a, 24). I agree with Hemdorff that this is a very plausible explanation, and it is also plausible that the loom-weights belonged to the house and were used here for textile production. Conversely, one could speculate whether the loom-weights could signify a break in tradition, as they were deposited at the end of the life cycle of the house. If so, they would represent a new habitus (see Bourdieu 1977, 1990 [1980])—a new way of being in the world. Noteworthy is that—as pointed out by Horve (2009, 63)—they point toward the advent of the secondary products revolution (discussed in chapter 5). In retrospect, now that type 4 hybrid houses are understood to be an architectonic reality, I am tempted to suggest that the postholes closer to each side of the walls in the western section of the house could form a trestle. If so, this house could also be a hybrid house. Olle Hem-

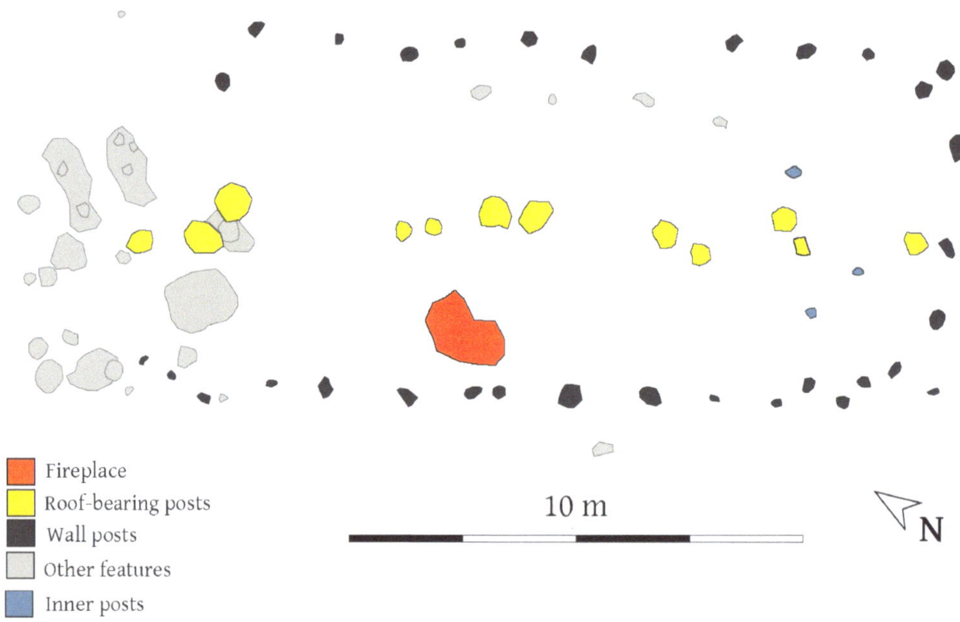

Figure 4.6 House 3 from Bjorhaug. Redrawn by Theo Bell Gil after Hilde Fyllingen (2015a, 30, fig. 29).

dorff, the excavator, communicated to me that he does not rule this interpretation out, but informed me that the postholes were shallow. The evidence is currently not strong enough to confirm that this is a type 4 hybrid house.

Bjorhaug

This site is situated at the flatlands of Hå, southern Jæren. Excavations on the two farms Bjorhaug and Bø disclosed agrarian settlement from the Early Bronze Age until the Late Iron Age. A farm consisting of a two-aisled longhouse, house 3, was established at today's Bjorhaug around 1600 BCE, followed by a break in habitation until the Pre-Roman Iron Age (Fyllingen 2015a). Altogether 12 roof bearing posts were found, representing different phases and bearing testimony to the replacement of posts, I suggest that 3–4 of these posts were standing at the same time. Traces of an internal wall in the southern part are seen in smaller postholes (see Figure 4.6). In the fireplace, a sherd of pottery with bone tempering was found, additionally, one other potsherd and a hammer stone were retrieved from the house. Grains from barley, emmer/spelt and wheat were found in postholes, as well as seeds from weeds (Fyllingen 2015a, 29–31).

Tjora

The farm of Tjora is situated on the Tananger peninsula, with the North Sea to the west and Hafrsfjord to the east. The name of the farm is by linguists considered to be pre-Norse, belonging to a cluster of names found along the coast of Jæren that might go back to the late Neolithic or the Bronze Age (Særheim 2011, 2015). The site was situated

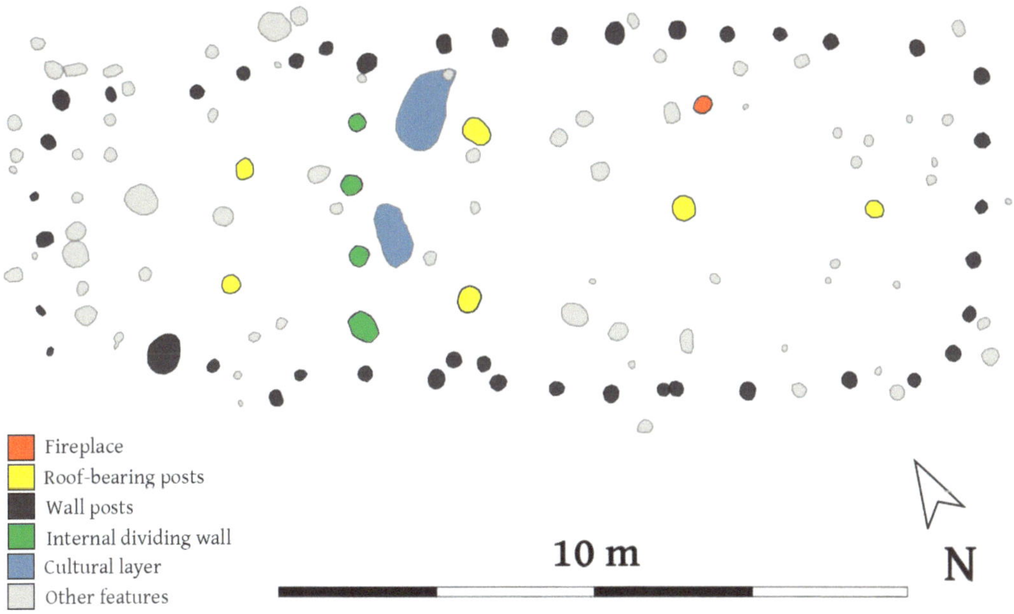

Figure 4.7 House 10 from Tjora, a hybrid house that combines two-aisled and three-aisled construction. Note the internal wall in the northern part of the house that separates the space between the two trestles. Features marked in grey are not construction elements of this house but could either belong to overlaying houses from other phases or be remains of activities that took place inside the house. Redrawn by Theo Bell Gil after Hilde Fyllingen and Niall John Armstrong (2012, 37, fig. 36).

near a high point in the landscape, Orshaugen, with a panorama view of the surrounding countryside.

The excavation of a large area uncovered remains of prehistoric settlements, graves and agricultural activities from the Late Neolithic until the Late Iron Age (Fyllingen 2015b; Fyllingen and Armstrong 2012). Altogether 17 houses were uncovered, ranging from the Late Neolithic to the Late Roman period. Here, six houses are considered as relevant based upon their dates. Of these, three are type 1, two-aisled longhouses (houses 9, 11 and 12), I interpret two as type 4, hybrids (houses 7 and 10), and one is type 3, three-aisled longhouse (house 4). The houses that are left out, is due to the lack of good dating or that only parts of the house was uncovered.

The two-aisled longhouses are well preserved with both roof-bearing posts and wall-posts intact, and the presence of several replaced roof-bearing posts attest to the long lifespans of the houses. The earliest of them is house 12, dated to the transition between the Late Neolithic and the Early Bronze Age. There might be an internal wall slightly north of the middle of the house, which could indicate a separation of space into different functions (Fyllingen and Armstrong 2012, 42–45). However, no other structures or structural differentiations give any clues as to which functions this might be. Adversely, houses 9 and 11 shows no markers of internal divisions (Fyllingen and Armstrong 2012, 33–35, 39–42). The four fireplaces spread evenly throughout house 9 indicate that it was living space for humans throughout. House 11 overlays house 9. Also this house shows no sign of internal divisions, or specialized use of space.

I interpret houses 7 and 10 as a type never recognized before in Norwegian archaeology, hybrid houses that combine the two-aisled and three-aisled construction, in which part of the house has a single row of roof-bearing posts, and the other part has a double row. Both are dated to the Early Bronze Age period I (Fyllingen and Armstrong 2012, 14; Appendix). A few instances of this type of house are also known from Jutland, from the sites Gunnerup and Fjordglimt, and are dated to Early Bronze Age period I, like the Tjora houses (Bech and Olsen 2013). The architectural outline of house 10 leaves no doubt that we are here dealing with a hybrid house (see Figure 4.7). An internal wall separates the northern part of the house into two inner rooms.

House 7 is less clear, partly due to a large boulder next to the western wall, which surely cramped this part of the house. Additionally, pits in the northwestern part of the house probably disturbed a roof-bearing post that would have paired up with the posthole in the northern corner and formed a trestle in the northwestern part of the house (see Figure 4.8).

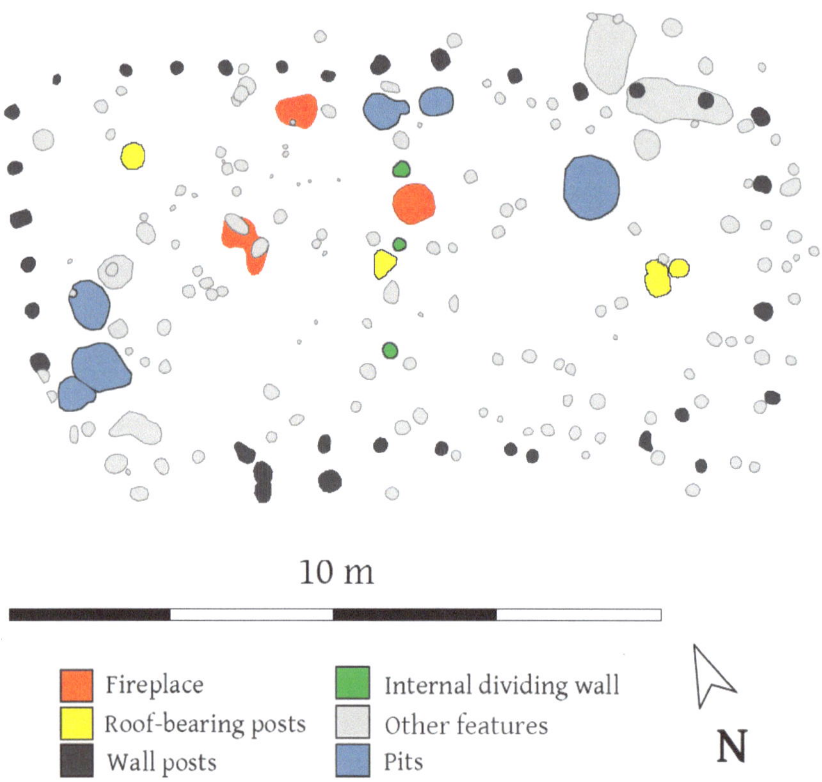

Figure 4.8 House 7 from Tjora. Note the three pits in the northwestern part of the house, where a roofbearing post might have stood. Features marked in grey are not construction elements of this house but could either belong to overlaying houses from other phases or be remains of activities that took place inside the house. Redrawn by Theo Bell Gil after after Hilde Fyllingen and Niall John Armstrong (2012, 29, fig. 24).

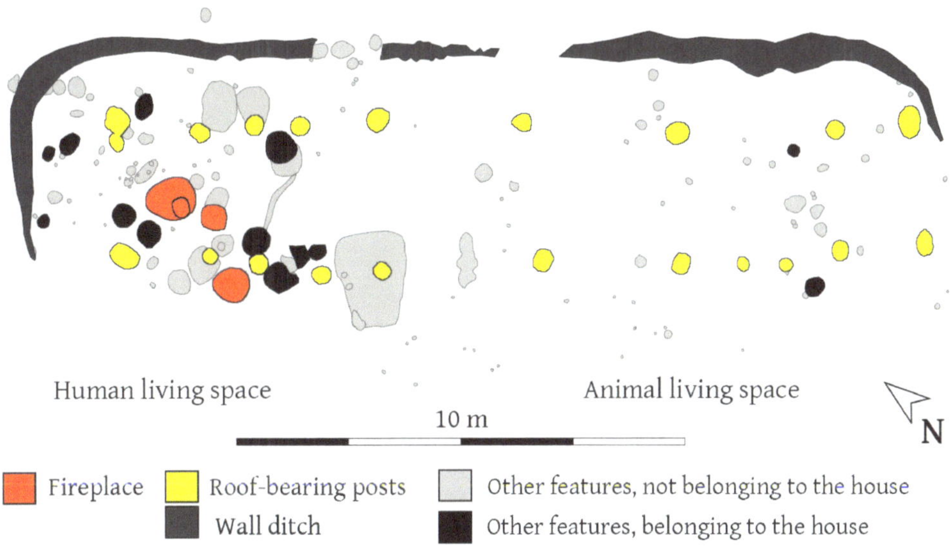

Figure 4.9 House 4, the earliest three-aisled longhouse at Tjora. Tentatively interpreted as divided into two parts, one for humans and one for animals. Features marked in grey are not construction elements of this house but could either belong to overlaying houses from other phases or be remains of activities that took place inside the house. Redrawn by Theo Bell Gil after Hilde Fyllingen and Niall John Armstrong (2012, 21, fig. 14).

Also in house 7 there is a clearly marked internal wall, which separates house 7 into two sections. However, the distribution of finds as well and other structures do not give any indication which functions the two parts had, and whether they had different functions. Remains of two fireplaces are present in the northern end, and one along the internal wall. The latter might not belong to the context of this house. Fragments of ceramics, flint (of these one microblade and one blade), struck rock crystal, burnt bone and an amber bead were distributed in the house, mainly in the northwestern section.

Both hybrid houses have relatively clear traces of postholes belonging to an internal, dividing wall. In two-aisled longhouses, such clearly marked internal walls are rarely seen (there are traces that indicate such walls but these are not clear), so the internal walls are therefore probably a novel trait.

House 4, see Figure 4.9, is the only three-aisled longhouse dated to the Early Bronze Age, and thus the earliest three-aisled longhouse on site (Fyllingen and Armstrong 2012, 20–24). It is well preserved, but lacks wall posts, possibly these were shallow. However, a ditch along the walls shows the full outline of most of the house. There is no clear internal wall, but several shallow postholes. A few of these, in a line along the middle, might have separated the house in two parts. Regardless, the house is clearly separated in two parts based upon the distribution of internal structures. Pits and fireplaces are found in the northern part. Several smaller postholes, making no apparent patterns, are found at both ends of the house. An entrance is seen as a break in the wall ditch towards the northeast, the other gap in the ditch further to the south is a result of the archaeological survey prior to the excavation proper. Thus, due to the lack of remains of wall posts and the lack of a wall ditch running along the southwestern long wall, it is uncertain if there were opposing entrances at this point, it seems more likely that there were separate entrances to the two parts.

During excavation, the house was interpreted as having two functions, the northern part with its pits and fireplaces as a living space for humans, and the southern part as a byre. However, phosphate analysis was done to test this hypothesis, but the analysis showed no differentiation between the parts, except an area with higher phosphate values outside of the eastern wall (Fyllingen and Armstrong 2012, 23). But phosphate mapping is by no means a reliable method. First, there is heavy post-occupation activity to be considered. The site has been farmland more or less continuously since the Early Bronze Age, alternating as pasture and for growing crops, which has greatly altered the soil chemistry over time. Grazing by animals creates a phosphate cycle, in which the animals take up phosphate through the grass, and then redeposit it (elsewhere) as dung, which tends to reduce the initial habitation patterns. Also, other natural processes by plants and animals contribute to relocation of phosphate (Barham and Macphall 2009, 95). Additionally, in the last 70 years all farmland on Jæren has been heavily exposed to industrial fertilizing, which has greatly altered the soil chemistry. Second, phosphate concentrations—or lack of them, cannot be dated. So many events have taken place at this particular site since the Early Bronze Age, and results from phosphate mapping could mark either one—or none—of them. Therefore, I do not trust phosphate mapping of prehistoric sites as a source of systematic knowledge of the past, since there are such great uncertainties inherent in its results. Rather, I believe that the initial interpretation is the most plausible one, and tentatively understand the house as separated in two parts, one for human habitation, and one for animals. On the basis of the distribution of structures, archaeologist Barbro Dahl supports this interpretation and disregards the phosphate mapping (Dahl 2014, 54).

Tjora is an important site to understand the transition from two-aisled to three-aisled houses, with its type 1, 3 and 4 houses. Noteworthy is that the two-aisled houses were preserved better with more internal structures than what is common (compare, for example, with Kvåle below), which gives a better insight into the use of internal space. The type 4 hybrid house(s) are a novel feature in Norwegian archaeology. In 2016, during an excavation at Tanangervegen, not far from Tjora, another house was excavated which is tentatively interpreted as a hybrid house, although not fully analyzed yet (Eilertsen In prep).[7] Now that this type of house has been identified in Norway and Denmark, excavators might be better prepared to recognize them in the future. Also, the type 3 house with its clear internal division based upon the distribution of internal structures, is important in comparison with the other early three-aisled longhouses.

Kvåle

The site of Kvåle on Jæren is unique considering its continued settlement from the late Neolithic until Medieval times, which has been studied in tandem with an extensive investigation of the well-preserved fossil landscape in the immediate vicinity (Soltvedt

7. Personal communication from the project leader, Krister Scheie Eilertsen.

et al. 2007). A permanent farm represented by a sequence of houses was established on one of the lower hills around 1900–1800 BCE, and remained at the same location for 400–500 years (Figure 4.10; Soltvedt *et al.* 2007, 45, 196). The oldest house (House 2) was a two-aisled longhouse. House 2 was replaced by a second, larger, two-aisled longhouse (House 1) dated to 1820–1720 BCE (Soltvedt *et al.* 2007, 39) built on top of House 1, this was a larger building that was oriented slightly differently. Subsequently, it was

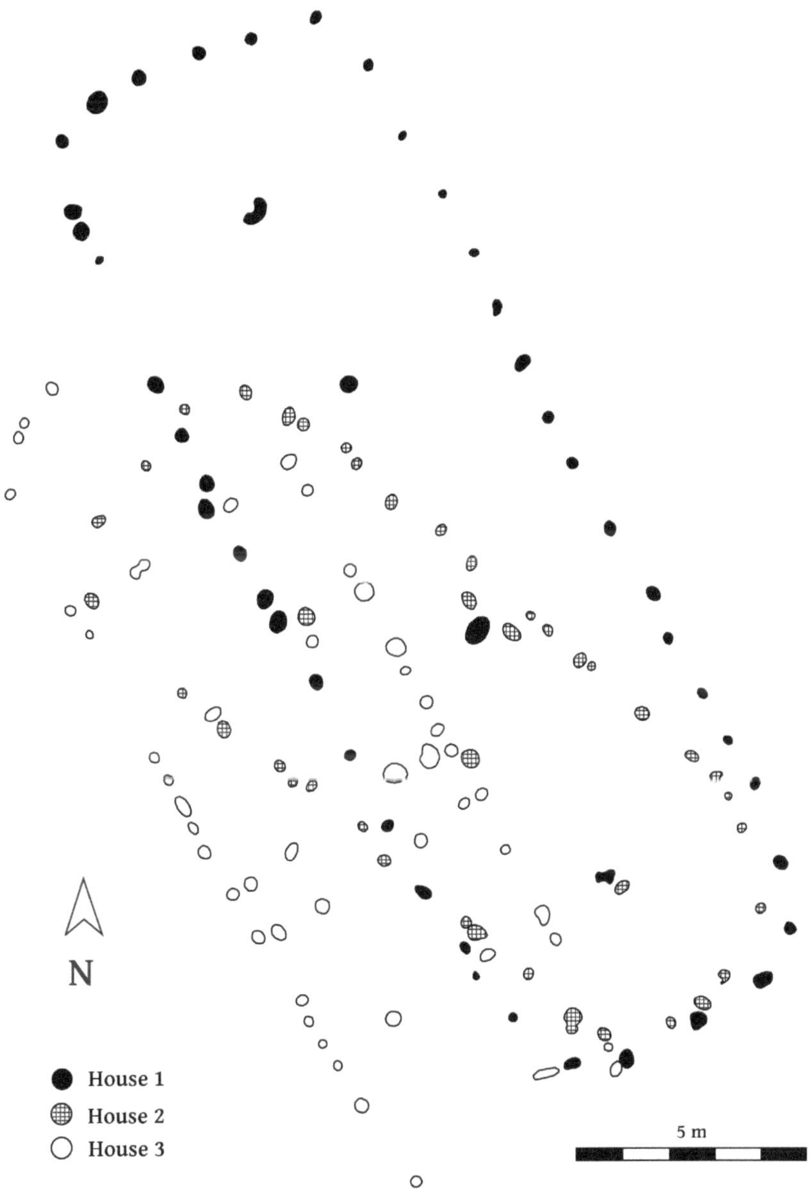

Figure 4.10 Houses 1, 2 and 3 from Kvåle. Figure by Ragnar L. Børsheim, Museum of Archaeology, University of Stavanger.

replaced by a three-aisled longhouse (House 3) dated to 1750–1665 BCE, which followed the orientation of House 1 and partly overlapped it. This is a unique situation since it captures the transition from one type of architectural choice and building technique, to another.

Both House 3 and House 2 had clearly marked opposite entrances at the middle of the house. There are thus both differences and similarities between House 3 and the other two houses. The difference in internal architectural layout indicates that House 3 probably had different functions compared to the other, earlier, houses, of practical and/or symbolic use.

Due to the lack of finds on floor levels, the interpretation of the houses' functions rests therefore on structural architectural choices, as well as finds and macrofossils from postholes. The postholes from House 3 contained a very low number of macrofossils, and in addition a few pot sherds, burnt clay and some flint flakes (Soltvedt *et al.* 2007, 49). In the surrounding area similar types of finds were made, all consistent with the Early Bronze Age find complex. The macrofossils from House 3 are barley, emmer wheat, wheat and oat. A very low number of seeds from weeds were also founds. These were scattered evenly throughout the house, which according to Soltvedt *et al.* (2007, 49, 76, 197–198) does not support an interpretation of a byre part. However, in retrospect I believe the numbers are too low to be conclusive, and further, one would not expect to find charred remains of weeds in a byre where animals were fed hay. Hay would not leave such traces. I argue that the macrofossils therefore are inconclusive regarding the use of the house. However, the architecture is consistent with indoor stalling of domestic animals—in particular the construction of the opposed entrances that form two separate rooms—but the overlapping of the three houses and the presence of opposite entrances in House 2 and 3 indicate a sliding rather than an abrupt transition.

Forsandmoen

The site of Forsandmoen holds a very special place in Norwegian archaeology, with its status as the first, and so far only, prehistoric village that was occupied over a long time. A multitude of house remains that span two millennia have been excavated (Dahl 2009; Løken 1997, 1998b). It is well published, most notably by one of the original excavators, Trond Løken (see for example Løken 1987, 1988, 1989, 1997, 1998a, 1998b, 2003). His work has left a rich legacy that the archaeology of the West Country can dip into again and again.

The topography lends itself to a settlement confined by natural barriers—this tiny flatland area is nestled in a hook of land where fjords meet and mountains collide, as it is nudged in between steep mountains and deep fjords, on a plain where Høgsfjorden forks into Lysefjorden. The flat and fertile plain is ideal for farming, and the surrounding mountains makes it more sheltered than the wind-swept Jæren region. The Forsandmoen excavations demonstrated that the processes of changes in the development of settlements that are seen in Denmark and northern Europe also happened in Norway. This opened up for a wide range of implications, and rather than seeing the Bronze Age of Rogaland and Norway as non-existing, Forsandmoen put Rogaland on the map of the Bronze Age world in terms of Bronze Age settlement patterns and lifestyle.

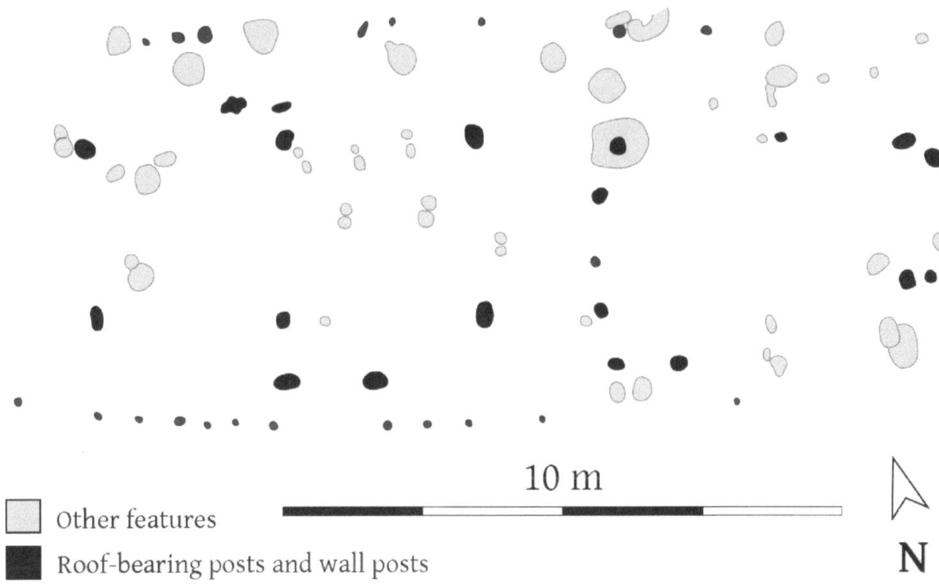

Figure 4.11 House XXXIIB (32B) from Forsandmoen. Features marked in grey are not construction elements of this house but could either belong to overlaying houses from other phases or be remains of activities that took place inside the house. Redrawn by Theo Bell Gil after Trond Løken (2003).

The settlement was seemingly established around 1500 BCE, Early Bronze Age II, and all houses are three-aisled. Altogether 254 house remains from this period demonstrate continuous settlement from c. 1500 BCE–600 CE (Løken 1997, 1998; Dahl 2009). The first phase of habitation is of special interest to this study, and will be considered here. Houses firmly dated to this period are XLV/45, XC/90, XXXIIB/32B, CCXII/212. House LIX/59 also have dates from this period, but additionally from later periods—but given that a good few of the dates are within the early range, it is considered here. The houses are described following the original report and drawings (Løken 2003).

House XXXIIB (32B), (Figure 4.11), was oriented WNW-ESE, and is the largest of the Bronze Age houses from Forsandmoen, with its size of 172 square m. The house is probably one phase of a series of buildings that have stood on the same location, it is therefore difficult to suggest that all of the posts that are found within the house actually belong to this phase. Three entrances are seen, two of these are somewhat asymmetrically opposed in the northwestern part of the house, and one is located along the southern wall, in the southeastern end. A clear division wall divides the house in two just north of the latter entrance. According to this division, the northwestern part is larger than the south-eastern part, 2/3 to 1/3, roughly. This division wall could indicate that the house was separated into a human living section and a byre. The considerable size of the house has led the excavator, Trond Løken, to suggest that it was, in fact, a hall (Løken 1998, 108–109).

A number of radiocarbon analyzes were done in order to establish which features belonged together, time-wise. The dates range between 1680–1400, making it the oldest

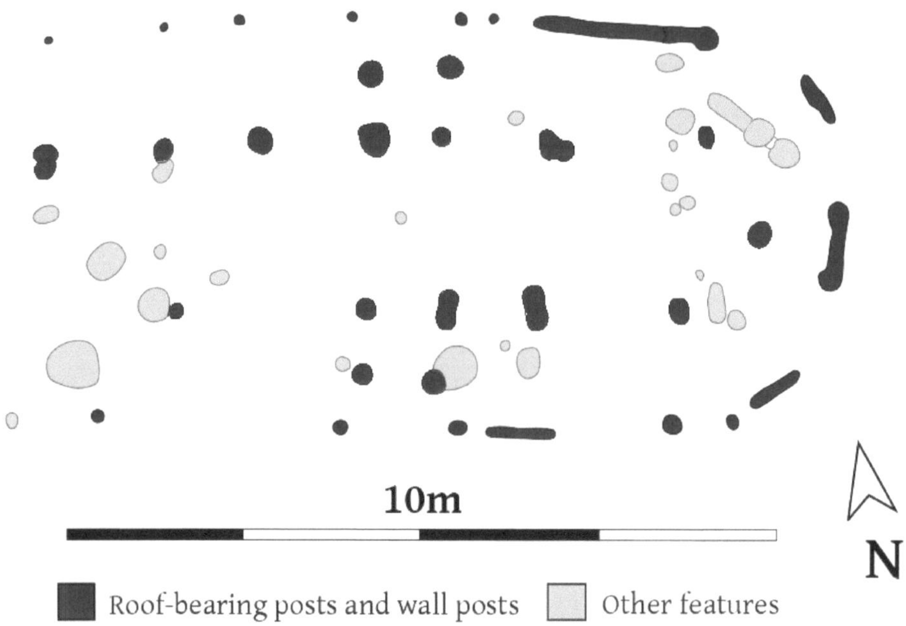

Figure 4.12 House XLV (45) from Forsand. Features marked in grey are not construction elements of this house but could either belong to overlaying houses from other phases or be remains of activities that took place inside the house. Redrawn by Theo Bell Gil after Trond Løken (2003).

house on site. The dates show that without a doubt, the interpretation of the house has consistency.

House XLV (45) was oriented WNW–ESE, and is medium sized with its 90 square m (Figure 4.12). Wall posts and wall ditches were preserved in the southeastern part of the house. Two entrances directly oppose each other at the mid section. In the southeastern part, some smaller postholes that could not be directly related to the construction of the house are seen. The excavators suggest that these could belong to another structure, but do not suggest what they might be. I suggest that rather than belonging to another structure, these postholes belong to the house. They could have been used to support posts that were falling down, although their positioning makes this unlikely, as they are clustered together. Or, they could be used for various "household functions" such as tripods to hang large cauldrons. One such structure could possibly be seen in the northwestern part of the house. However, the fact that the postholes in the southeastern part are clustered so closely together seems like a counter-argument. Rather, these posts might have been in use one or two at a time, pulled out, and in due time replaced by another at roughly the same position. One possible solution to this puzzle is that they indicate makeshift poles put down for tethering animals in the southeastern part of the house, or to make smaller makeshift pens.

Further, it seems remarkable that the southeastern part is better preserved than the northwestern part. One of the excavators, Trond Løken, communicated to me that this is a recurrent trait with the Bronze Age houses from the site, and his interpretation is

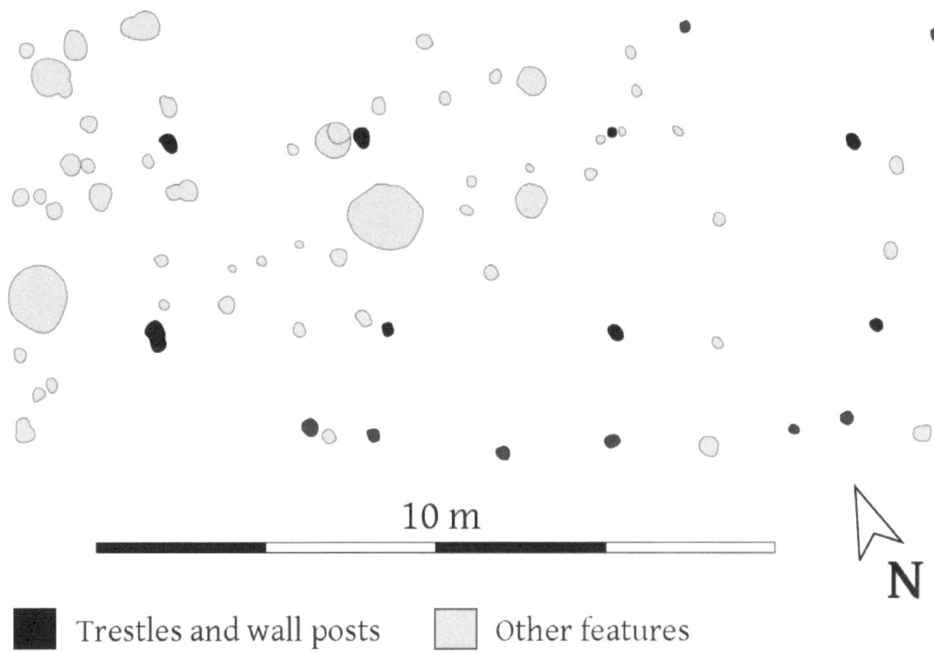

Figure 4.13 House XC (90) from Forsandmoen. Features marked in grey are not construction elements of this house but could either belong to overlaying houses from other phases or be remains of activities that took place inside the house. Redrawn by Theo Bell Gil after Trond Løken (2003).

that this part of the house was used for indoor stalling of animals. Thus, it became much more heavily trodden, leaving the floor level deeper and therefore also better preserved than the part of the house that was living section for humans exclusively. Considering that the postulated byre was better preserved, traces of stalls would be expected if cattle were kept inside, but there are no such traces. This indicates to me that cattle were not stalled here, and I suggest that instead, animals were kept in makeshift pens or tethered. Probably these were smaller domestic animals, such as sheep and/or pigs. This phenomena could also explain the clusters of smaller postholes in houses 10, 7 and 4 from Tjora, discussed above.

House XC (90) was also oriented WNW–ESE, and was medium sized, with its 116 square m (see Figure 4.13). Wall posts were preserved along the southern wall. Large amounts of charred grains were found in the roof-bearing posts, they were even visible on the surface. No entrances are seen. A multitude of smaller posts and features are seen, that presumably belong to another house. Information about this house is scant. It has been dated to 1500–1260 by 4 samples taken from four different postholes, showing great consistency.

These three houses, XLV, XC and XXXIIB, are located adjacent to each other, in the north-eastern part of the plain, situated a bit higher up and closer to the eastern mountains compared to how the habitation came to unfold in later periods. House LIX (described below) was also situated in this area.

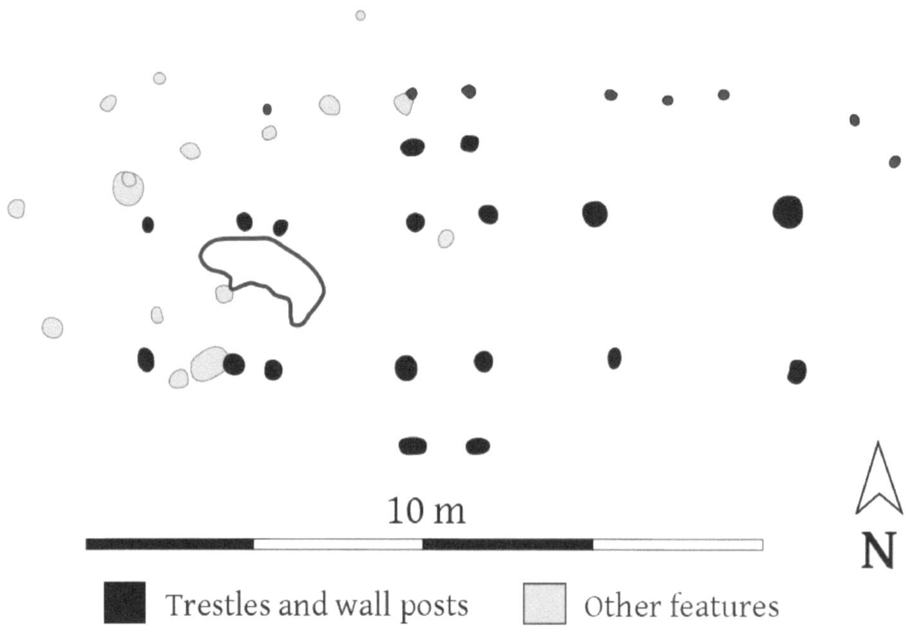

Figure 4.14 House CCXII (212) from Forsandmoen. Features marked in grey are not construction elements of this house but could either belong to overlaying houses from other phases or be remains of activities that took place inside the house. Redrawn by Theo Bell Gil after Trond Løken (2003).

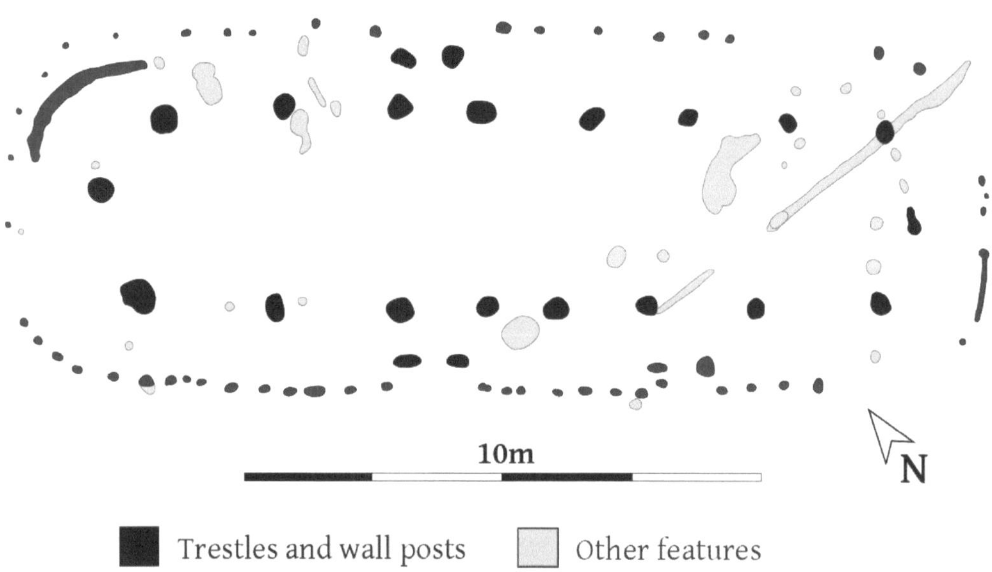

Figure 4.15 House LIX (59) from Forsandmoen. Features marked in grey are not construction elements of this house but could either belong to overlaying houses from other phases or be remains of activities that took place inside the house. Redrawn by Theo Bell Gil after Trond Løken (2003).

House CCXII (212) was oriented E–W, and with its 101 square m it was of medium size (see Figure 4.14). The wall posts along the northeastern part are preserved, and entrances are found facing each other at the middle. The trestles were placed more widely apart in the eastern section, both between the trestles themselves and from one pair of trestles to the next, making this part of the house fairly open. The excavator interprets the eastern part as the human living section, due to its spatial openness (Løken 2003). Further, a quantity of pottery was found in this part, as well as traces of a clay floor, seen in large amounts of clay in the roof-bearing posts. A radiocarbon date from one of the roof-bearing postholes of the Early Bronze Age III house dates to 1260–900. It is somewhat later than the three houses described above, and is situated further away, towards the southwest, centrally on the plain.

House LIX (59) has proved to be difficult to date accurately, since several radiocarbon samples from the house have given a wide range of dates from Bronze Age periods II–V (see Figure 4.15). However, I have included it in my dataset, since the dating allows for it to have a place in the earliest period of habitation, and the architectural layout is consistent with Early Bronze Age type 2 houses. It was oriented NW–SE, and was comparatively well preserved. The wall posts were preserved, and some places also ditches along the wall. The walls had rounded corners. There were three entrances, two facing each other in the middle/northwestern part, and one further south along the southern wall. The opposing entrances seem to mark a partition (see also Løken 2003), the southeastern part has more densely spaced roof-bearing posts, both regarding distance between posts in one trestle, and between the trestles as pairs. It is also in this part that there is a third entrance. I interpret this part as a byre, considering the more densely spaced trestles. Also, smaller posts seen in this part could have been used for tethering or makeshift pens. The northwestern part would most likely be the human habitation part, considering its more spacious architectural solution. No fireplaces have been found.

Finds of flint tools produced by pressure flaking confirms a Bronze Age date. There are also copious amounts of pottery from this house. Large amounts of grains were also retrieved from the postholes.

Myklebust

Myklebust is situated close to the sea at the northern part of the Tananger peninsula, a few kilometres north of Tjora. It is close to the monumental mound Sothaug, which is also dated to the Early Bronze Age period II (see chapter 6). At the site, altogether seven houses from the Early Bronze Age were excavated (Dahl 2014). Three of these were longhouses, and only one, house XV (15), is a "classical" longhouse. House VIII is included here to show diversity, although it barely fits the criteria of a longhouse as it is very different from most of them. The remaining houses were small four-post constructions. House X is not included, both due to its later date, to the transition between Bronze Age III–IV, and due to the uncertainties surrounding the interpretation of the house(s) (Dahl 2014, 64–65).

House XV (Dahl 2014, 49–55) is a longhouse of type 3 with five trestles, or possibly six (see Figure 4.16). In the eastern section, several fireplaces and pits were found, whereas one pit interpreted as a storage pit was situated in the northern corner of the western

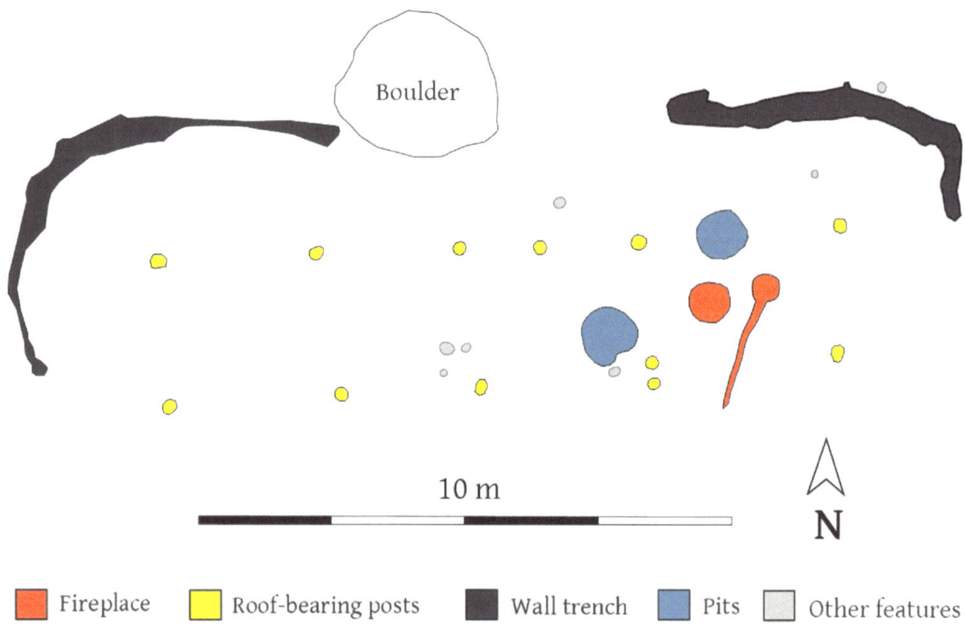

Figure 4.16 Myklebust house XV (15). Redrawn by Theo Bell Gil after Barbro Dahl (2014, 50).

section. A ditch connected one of the fireplaces in the eastern section to the outside of the house, interpreted as an air vent.

Slightly west of the middle two smaller postholes are interpreted as the remains of an internal wall which separated the house into two parts: living space for humans in the eastern section and a byre in the western section (Dahl 2014, 49, 72). Barbro Dahl, the archaeologist in charge of the excavation, compares the house to houses excavated at Forsandmoen and suggests on the basis of the internal separation that the house had opposing entrances at the middle. Or, the entrances could be similar to House 4 from Tjora (see above), with separate entrances for the two sections. These houses are remarkably similar regarding the distribution of pits and fireplaces in one section, and also the wall ditch found along one side of the house (Dahl 2014, 54). Dahl interprets House 4 from Tjora and House XV from Myklebust as longhouses divided into two sections, one byre and one human living section.

House VIII is a very small three-aisled longhouse, just barely bigger than the four four-post buildings excavated at the site (Dahl 2014, 43–48). The spacing between the trestles varies between 1–3.5 m. A wall-ditch surrounds one long wall and one short wall, and posts interpreted as markers of an entrance is seen adjacent to the ditch. The ditch continues also where the entrance is thought to be. Nothing indicates that there were internal separations of space into different functions and Dahl interprets it as a human living-space (Dahl 2014, 48).

Lunde

Lunde is found on the northeastern part of Hundvåg, an island just off the town of Stavanger, at the northern part of Jæren. It is fairly close to Austbø, which is discussed below. House 1 and House 3 from Lunde are both dated to the Bronze Age. House 1 had three trestles, with posts that had clearly been replaced, attesting to its relatively long lifespan (Pedersen 2013, 24–29). It is similar to the houses at Myklebust, both regarding size (it is small), the presence of a wall ditch that surrounds the eastern short wall, and a fireplace in the eastern part with an air vent. The size of the house precludes an internal division into differentiated functions, and I interpret the house as used primarily for human habitation. There is nothing to indicate that a part of the house was a byre (Pedersen 2013, 42). House 3 is not complete and is therefore not considered here. It was clearly a small house, similar to House 1.

Kleppestemmen

At Kleppestemmen, two houses were excavated (see Berge 2007). House II was a two-aisled, late Neolithic longhouse oriented S–N. Adjacent to the north, but oriented E–W, was House I, a three-aisled longhouse dated to Early Bronze Age II (see Figure 4.17). The excavator interprets the two houses in relation to each other, and suggests that house I preceded house II as a dwelling (Berge 2007, 16). However, the time gap between the houses is considerable, from the Late Neolithic to Bronze Age period II. The excavators found no traces of entrances or other architectural features. However, I note that the trestles are situated more closely together in the eastern part of the house. Therefore, the house is tentatively interpreted as divided into two sections, with a human living section and a byre.

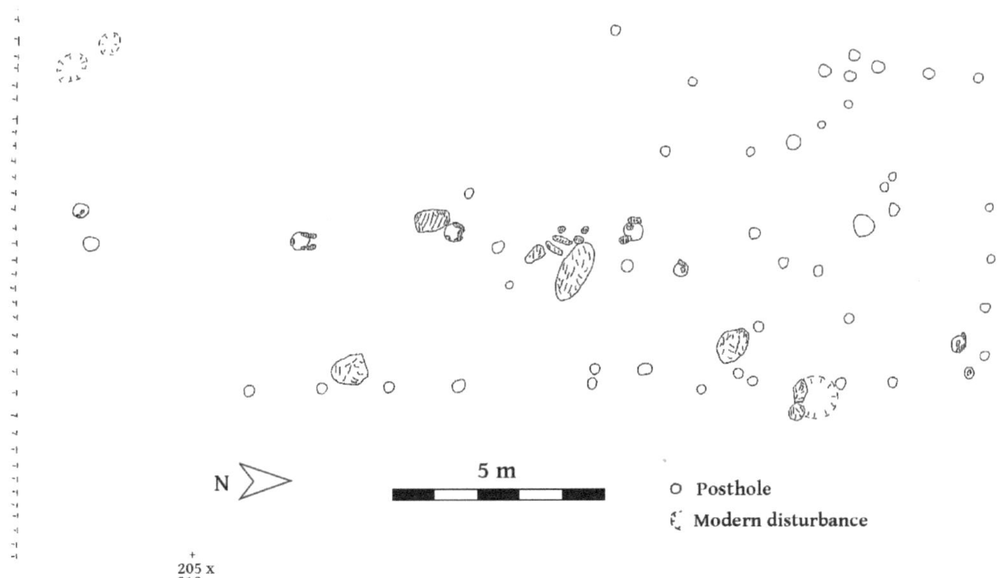

Figure 4.17 Kleppestemmen house I. Note the number of smaller posts in the northern section. After Jan Berge (2007).

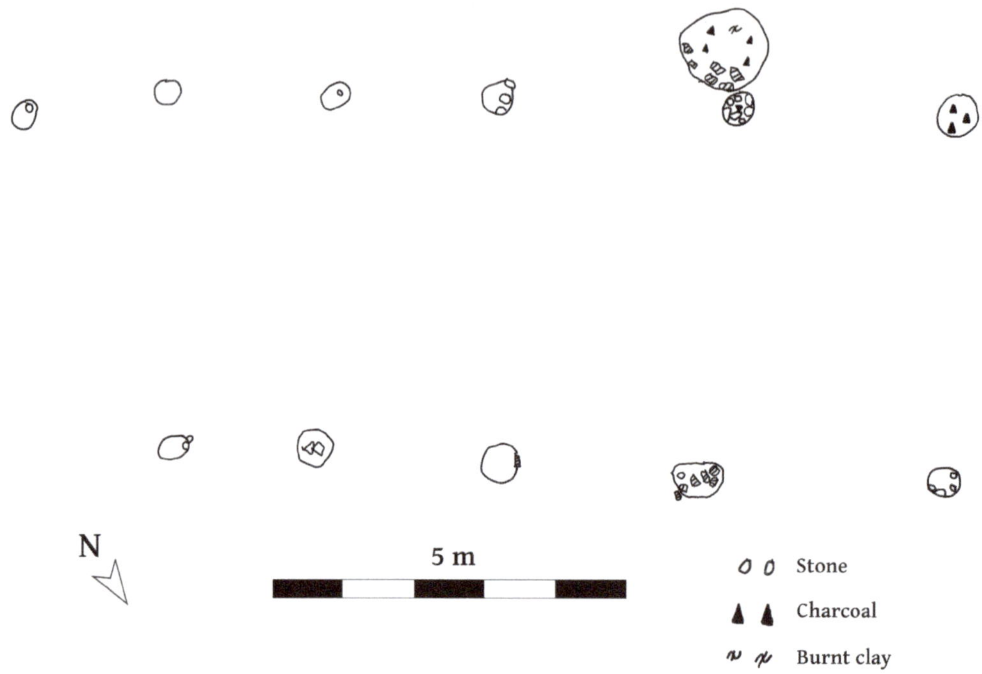

Figure 4.18 Skeiane house 1, modified by Inger Horve (2009, 84) after the original by Frode Pilskog (1998b).

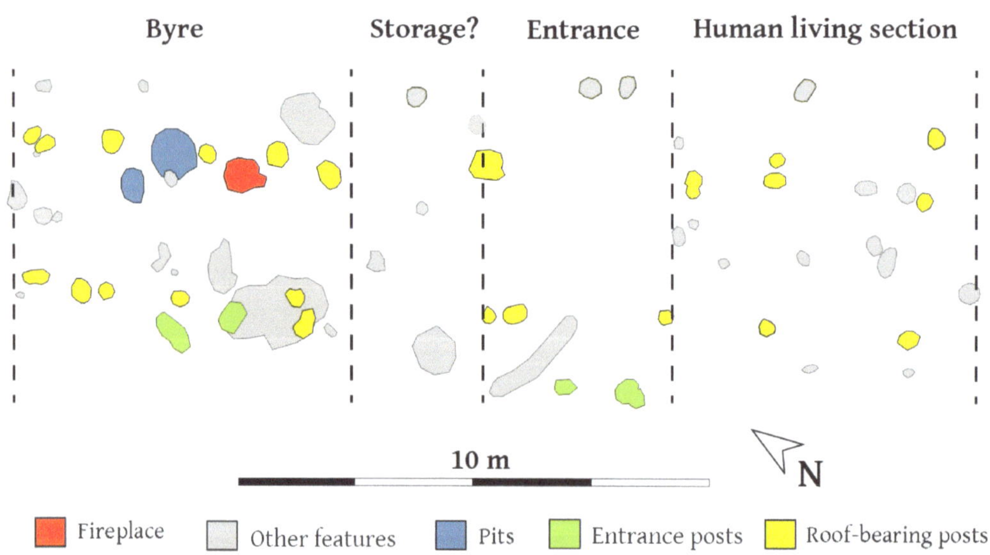

Figure 4.19 House 3 from Tengesdal with all internal structures. Note that the trestles in the northeastern section are more densely placed. Redrawn by Theo Bell Gil after Kristine Reiersen (2015, 67, fig. 50).

Very few finds were made on site, some pieces of pottery, flint and burnt bone. In addition, macrofossils were taken from the structures. These contained high numbers of charred grains. In house 1, burnt bones thought to be animal bones were found in the three roof-bearing posts in the northern part of the house.

Noteworthy is the situation where a two-aisled house seems to directly precede a three-aisled house. We are here able to catch a glimpse of the architectural transition. Interestingly, like at Kvåle, the three-aisled longhouse changed direction in relation to the two-aisled longhouse, as though the builders wanted to signify the break in tradition that would be seen also from the outside, by travellers far away, even.

Skeiane

House 1 from Skeiane, Sandved (see Figure 4.18), has not been dated, but is typologically dated to the Early Bronze Age, based upon the structural similarity with houses from Forsand that are dated to this period (Pilskog 1998a, 13; see also Pilskog 1998b). In addition a storage pit A324 is dated to 1265–925 BCE (Pilskog 1998b, 28). Presumably, this pit belonged to the house and was used together with it. Other than the trestles and the storage pit, other features were not found, such as walls and entrances.

The house is built in sloping terrain, and the trestles at the lower end are placed more densely together, and were also more shallowly established, or dug down. Although there are no structural traces of a byre, no stalls and also no signs of entrances, both the more densely placed trestles and the a more shallow foundation at the southeastern part could indicate greater wear and thus that domestic animals were kept in the lower part of the house.

Tengesdal

The site of Tengesdal is the southernmost site considered here, in the region of Dalane (The Valleys) south of Jæren. It is situated on a flat plain close by the Bjerkreim river, in a region that is more hilly compared to the flatlands of Jæren. The site comprised the remains of at least three prehistoric longhouses as well as other structures, such as fireplaces and cooking pits, as well as a layer of undisturbed prehistoric agricultural soil (Reiersen 2015). One of the three-aisled houses was dated to the Early Bronze Age, but unfortunately a two-aisled house was not dated.

House 3 is a longhouse of type 3 and is dated to Bronze Age period II based on macro fossils from postholes and one of the fireplaces (see Figure 4.19; Reiersen 2015, 41–49). In the northwestern part, the trestles seem like they are placed more closely and there are some smaller postholes in this area. However, some of the trestles look like they have been repaired, so presumably not all of the posts indicated by the postholes have been standing at the same time. Three fireplaces are also found in this part, one of them is dated, and is slightly later than the dates of the postholes (Reiersen 2015, 48). The placement of the fireplaces near the northeastern row of postholes makes it unlikely that they belong to the house. Smaller postholes are also scattered in the southeastern part. Two entrances are found along the southern wall, one in the northwestern section and one in the southeastern. An interesting feature of this house that gleans insight into the use of internal space is the distribution of macrofossils found in the postholes.

In the northwestern part of the house, the roofbearing posts contained seeds from weeds, such as meadow plants and grass from pasture. This could be consistent with indoors foddering of animals and haymaking. In the southeastern section, grains were found in the postholes. Based upon this, Reiersen tentatively interprets the northwestern section as a byre, the middle sections as entrance and storage, and the southeastern section as human living-space (Reiersen 2015, 65–67).

Austbø

Austbø is situated in the southwestern part of the island of Hundvåg, close to the sea, and not far from Lunde. Several sites—4, 19, 20, 21 and 22—are dated to the early Bronze Age and seem to belong to the same complex—one farm (Juhl 2001, 42–51, see map fig.

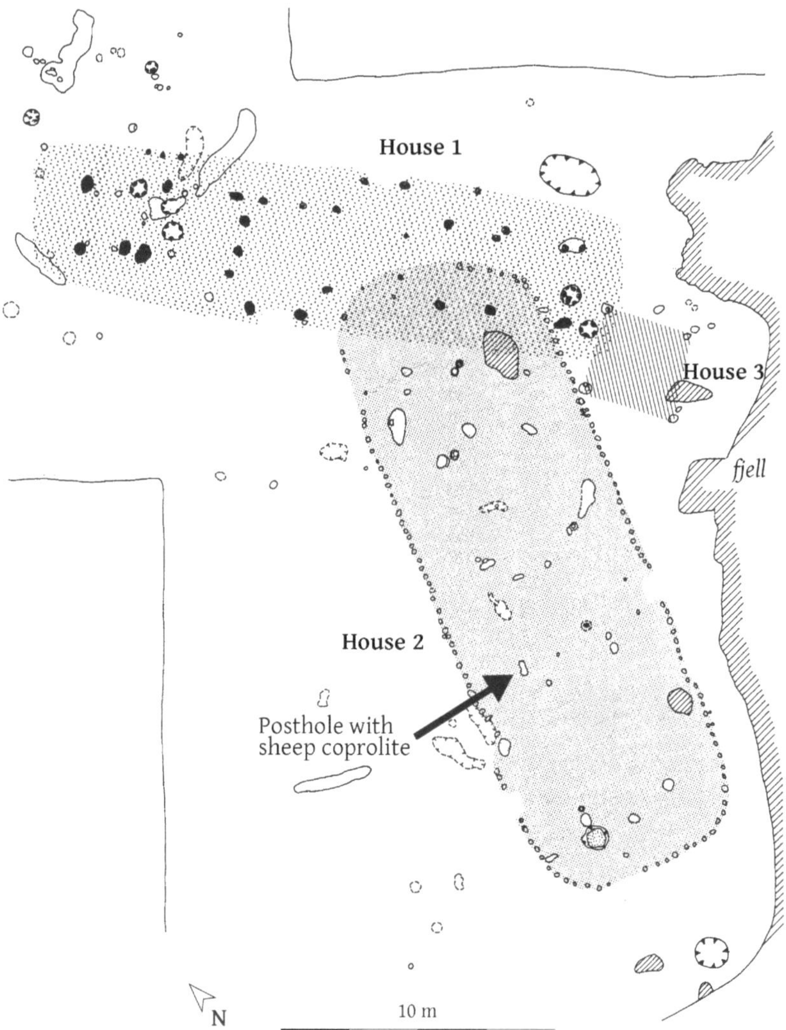

Figure 4.20 Austbø house 1 (top) and house 2. Sheep coprolite found inside posthole AG. Redrawn after Kirsten Juhl (2001, 46).

23, p. 42). The sites probably reflect a variety of activities although what these are cannot always be determined with certainty. One of the sites (19) is a rock art panel with 11 cupmarks, presumably made by the people living there in the Bronze Age. Two sites (4 and 22) are large midden-like sites with extensive cultural layers but no traces of constructions.

On the site of Austbø lok 20, two three-aisled houses from the Bronze Age were excavated (see Figure 4.20). These are both dated to Early Bronze Age II, and are early three-aisled longhouses. House 1 and 2 cannot have been contemporary, since the northern part of House 2 and southwestern part of House 1 overlap. Archaeologist Kirsten Juhl (2001, 45–48) tried to solve the mystery of which is the older house. The dates are inconclusive but House 2 could be older. This indicates that neither house had a very long lifespan, but possibly a lifespan of 120–140 years (Juhl 2001, 48). Hemdorff (1989, 6) reaches a different conclusion in his original report, he reckons that house 1 is older based upon the fact that a large amount of charred organic material was found in house 2, and he would expect to see some of that material ending up in the postholes in house 1 if this was older—but that has not happened.

House 1 has a remarkable outline, it is very similar to Forsand house XXXIIB as well as a house from Trappendal in Thy, northern Jutland, with a byre part in the middle of the house (Hemdorff 1989, 4). Traces of stalls have not been recovered from the Austbø 1 house, but some smaller posts along the northern long side of the house indicate that stalls were possibly placed there—or that the posts were used for tethering animals or make-shift pens. Further, the entrances are placed in the middle "byre" section. As pointed out by the excavator, Olle Hemmdorf, it is more practical to have the people walking through the animal part than the other way around.

House 2 is of the same size as House 1, but with a slightly different outline. There are fewer trestles, the posts are spread more widely, and it seems to be separated into two parts by two closely set posts across in the middle of the house, these are supposedly doorposts for an opening between the two parts. There are two entrances, set asymmetrically at each long side. The postholes contained much charred material, notably in some cases copious amounts of charred grains, which led the excavator to believe that the house must have burnt at least once (Hemdorff 1989, 8). Remarkably, charred sheep dung was found in the posthole dubbed AG, the posthole to the south of the second trestle from the west (see Figure 4.20). The post had been replaced (Hemdorff 1989, 8), so the sheep dropping, found in the very bottom of the posthole (Hemdorff 1989, 11), must have fallen into the hole as the post was replaced, i.e. not during the intial construction of the house, but some time during the lifespan of the standing house. This means that sheep were inside the house, and some quantity of dung must have been present for it to end up at the bottom of a posthole. I believe this demonstrates that sheep were indeed living inside the house. After all, sheep coprolites are a by-product of living animals.

Horve (2009) did not believe that animals could have been kept inside the house due to a feature in the southwestern corner—a circular slate-slab interpreted as a slab for either baking bread or drying grains. However, in the published results from the excavation, Juhl (2001, 43) gives a much earlier date for this feature, it is dated to the Neolithic and therefore not associated with the house and the Early Bronze Age settlement.

Considering the evidence, I find no objections to suggesting that sheep lived inside the house. Probably their life-space was the southeastern part of the house, based upon the fact that the house seems to have been divided into two parts, one southeastern and one northwestern, and sheep droppings were found in a posthole in the southeastern part. The droppings must have been deposited some time after the house was built, and ended up in the posthole during repair work.

Another probable sheep dropping was found in either a pit just outside the house or in one of the wall posts in the northeastern corner of the house (the report is not altogether clear on the exact location, see Hemdorff 1989, Appendix 3, find no bc58). I suggest that this ended up there because there were sheep in the area that were familiar with both the inside of the western part of the house and were intimate with the farm and its surroundings.

Finding houses shared by humans and animals

Before discussing the evidence from the Rogaland houses, let us take a closer look at factors that are expected to be present beyond the basic layout when hunting for byres. Waterbolk (1975, 385) has established the following criteria for indoor stalling of animals in three-aisled longhouses:

- Parallell dividing walls across the long sides of the house
- Extra posts at the inside, and in line with, roof-bearing posts (indication of stalls)
- Ditches in the lengthwise direction
- Flooring of stone or wattle
- Roofbearing posts more densely placed in the byre section than in the human living section.

However, we see from Tesch's and Horve's typologies outlined above that Scandinavian Bronze Age longhouses show great variation. Byres have not been identified in all longhouses and only a few especially well preserved examples have clearly defined stalls (Rasmussen 1999, fig. 1, p. 282). It should be considered that excavations of settlement areas are often ridden by a preservation bias, since the method used is topsoil stripping. The top layers are mechanically removed, which leaves sub-soil features, postholes and pits, these might contain finds such as objects, bones and macrofossils. Floor levels and find clusters are very rarely found. But, the previously mentioned recently excavated houses by the Limfjord on Jutland have, by their excellent preservation, disproved the attempt by Waterbolk to create neat criteria. Altogether seven such houses have been unearthed—houses that have burnt with humans and animals inside, and the remains from the various individuals were found where they perished in the fire (Kveiborg 2008, 2009; Nielsen 2002; Olsen 2007). None of these houses were constructed in accord with Waterbolk's criteria. For example, no stalls were found in the house from Søndre Ydby, even though cattle, horses and a sheep were housed in the byre part. This demonstrates that there is greater complexity and several possible solutions when it comes to keeping animals in the house. On these grounds I therefore disregard that such strict criteria as those listed by

Figure 4.21 Example of a Bronze Age three-aisled longhouse with a byre section, where animals are presumed to have lived in the one part, and humans in the other. Illustration: Ragnar L. Børsheim / Arkikon.

Waterbolk must necessarily be present when considering the nature of inhabitants in the three-aisled longhouse (Figure 4.21). This opens up for a broader perspective of shared housing between humans and animals. There are numerous ways of keeping animals indoors.

Evidence for indoor byres is frequently inferred from the occurrence of visible traces of stalls. None such are seen in the Early Bronze Age houses in Rogaland. But what do stalls indicate? They indicate assigning larger domestic mammals, such as cattle and horses, to stalls. Frequently, the discourse around bringing animals into houses in the Early Bronze Age uses the generic term animal, but—as discussed above—what is actually referred to is cattle. This demonstrates that we must be wary of collapsing the "animal" category to tacitly mean cattle, and take into consideration the multitude of beings that animals, as a group of sentient creatures, make up.

Did animals live in the Early Bronze Age houses of Rogaland?

In order to understand whether—and if so, in what manner—domestic animals shared the longhouse with humans in the Early Bronze Age in Rogaland, I will briefly summarize the findings of the study of the houses.

- First, common to all houses is a tendency towards a northwest/southeast orientation, both for two-aisled and three-aisled houses. Five of the houses are oriented west/east, one of these is two-aisled (Talgje). One of the two-aisled houses (Tjora house 12) is oriented north-northeast/south-southwest, whereas one three-aisled house (Austbø house 2) is oriented north/south. This remarkable consistency is probably a pointer towards the preservation of some traditions across the transition.

- Second, according to the evidence at hand, there are no indications of two-aisled houses having a byre-part. A few of the houses might have had an internal wall separating the house into two parts, but there is nothing to indicate that either of the ensuing sections were byres.
- Third, of the altogether 16 three-aisled houses, 11 are interpreted as having a byre, Kvåle house 3, Kleppestemmen house II, Austbø houses 1 and 2, Myklebust house XV, Tjora house 4, Forsand houses XLV/45, XXXIIB/32B, CCXII/212, LIX/59 and Tengesdal house 3. This is based upon the distribution of inner structures, entrances, dividing walls, more closely spaced trestles in one section, smaller posts interpreted as being used for tethering or makeshift pens and in one case the distribution of macrofossils. Not only I but also the excavators have in most of these cases interpreted the houses as divided into a human living section and a byre. Some of the houses (Kleppevarden house 1, Forsand house XV/90, Skeiane house 1) are not sufficiently well preserved to tentatively suggest that they had a byre. And finally, some houses probably didn't have byres (Myklebust house VIII and Lunde house 1). There is no consistency regarding types 2 and 3, both types have several houses that are interpreted as byres.
- Fourth, common to all of them is that there is no indication of stalls inside the houses.
- Fifth, there is a great variety between the houses, and a great inner variety within the postulated types 2 and 3. Types 1 and 3 are the most common kinds of houses, type 4 is so far very rare but not an anomaly. I predict that more type 4 houses will come to light in the coming years.
- Sixth, evidence is so far too scant to suggest the functions of house type 4. Both of the houses considered here seem to have internal walls dividing the house into two sections, but beyond that it is hard to tell which purposes the sections served. Presumably, this transition type is a form of experimentation. Animals could have been kept in one part, but if so, possibly not consistently, and probably not in a manner that would leave traces. Keeping animals inside was probably connected with particular parts of the year with greater vulnerability on the part of the animals, and such a use might have been seasonal. For example, related to lambing and the early spring, see chapter 5.

My understanding of the houses is therefore in opposition to Horve (2009, 44), who states that in Early Bronze Age Rogaland, there are no houses as yet that have without a shadow of a doubt housed humans and animals both. This is based upon the absence of clearly defined stalls, and the scant evidence of more densely placed posts in the postulated byre section of the house.

However, my dataset includes new material, which includes some well-preserved houses that are informative regarding the distribution of internal structures. Also, I place less weight on the presence or absence of byres. The houses from Early Bronze Age Rogaland might have had byre parts, indicated by the opposing entrances, but there are no internal divisions that indicate the presence of stalls. Perhaps this is due to a preservation bias, but we cannot—due to potential biases—conjure up archaeological data that does not, in fact, exist! Rather than thinking of these houses as lacking stalls, we should see of them as never having been constructed with stalls in mind. Thus, it seems unlikely that animals were living in byres—with clearly defined stalls—inside the houses. But this does not mean that animals did not live inside in other ways. For example, smaller domesticates such as sheep would not be tethered in stalls, but rather be free-range and at special times such as in the lambing season live in makeshift pens. In a practical sense, a three-aisled construction is an obvious advantage in a sheep pen, and well suited to accommodate sheep, since the distribution of posts makes it easier to set up makeshift pens. The fact that the posts are placed more densely in one part of the house indicates to me that they were used as supports for makeshift pens. Further, this means that animals might not have lived inside all year round, but rather during particular seasons, like the spring. At other times of the year, the byre might have been used for other purposes or activities, such as for craftwork or for storage.

Considering the obtainable evidence of three-aisled houses in Rogaland, what inferences can be made about who were members of the household? No traces of stalls are seen, however, the internal division of space frequently shows two or more parts with different architectural solutions. This is the case with Forsandmoen houses CCXII (212) and LIX (59), Austbø house 1 as well as Skeiane house 1, Tengesdal house 3, Tjora house 4 as well as the two hybrid houses 10 and 7, Myklebust house XV and possibly Kleppestemmen house II. As suggested above, the built structure embodied in the three-aisled longhouse is in itself indicative of human and animal household members. Particularly the construction of opposing entrances in the middle of the house, dividing the house in two, is consistent with indoor stalling of domestic animals. This trait is seen in Kvåle house 3, Austbø house 1, as well as Forsandmoen houses XLV (45), CCXII (212) and LIX (59). But also asymmetrically opposing entrances could indicate indoor stalling of animals. Both entrances of the Austbø houses are asymmetrically opposed (even though the entrances in House I leads into the middle), and seem to be entries into different sections of the house. Tengesdal house 3 had separate entrances into different sections, seemingly also a solution that was practiced regarding byres. Another factor that has frequently been overlooked are smaller postholes that are placed "randomly" inside the house and are not part of the construction of the house. Such posts could represent tethering of animals or temporary, makeshift pens. We find such postholes most clearly in Forsandmoen house XLV (45), XXXIIB (32B), XC (90), LIX (59), Tjora house 4 and hybrid house 10, Kleppestemmen house II and to a certain degree in Austbø house 1.

An inherent bias regarding such small post-, or pole-holes is that they are not always considered important and particularly in older excavations, they might not have been documented. Looking at the original documentation of the Forsand excavation made evident to me that even though they are documented, they often do not make it into the polished, published version that is generally accessible (see, for example, Figure 4.12 and Figure 4.13—the features marked in grey are from the original drawings but are not included in the publications, thus the distribution of smaller, internal posts would not become public knowledge). Without access to the original documentation I would not have been able to study, or even be aware of, these postholes, and certainly not come to any interesting, or otherwise, theories regarding them. Another problem is that at complex settlement sites, it might not be possible to tell which houses, or contexts, they belong to—if any. Good case studies are needed. For example, house 4 at Tjora (see Figure 4.9) was placed apart from other structures, and it was obvious that the smaller posts were clustered inside the house, hardly any were found outside of it (personal communication Niall Armstrong).

Such source criticism is also valid for two-aisled houses. For example, in Bjorhaug House 3 (see Figure 4.6), there is a row of smaller posts inside of the northeastern wall posts, and also a few smaller posts in the southern section. What kind of practices do these represent? Could they be indicative of early experiments with penning animals inside houses? Here, I will briefly mention two houses that have not been included in the data set: Kvia and Jåsund. These two-aisled longhouses have both been examined rigorously to look for functional division into living part for humans and byre. Kvia is dated to the Late Neolithic (Bjørlo 2011), and thus falls outside of my selection criteria. The dating of Jåsund is very broad, from 2300–1600 BCE, placing it in the Late Neolithic and beginning of Early Bronze Age period I. It appears to have had two phases (Fyllingen 2012, 111). Both of these houses have traces of an inner, dividing wall. The distribution of macrofossils is not even throughout the house from Jåsund, charred grains (but very few weeds) are found in greater numbers in the northern section, particularly along the short wall (Fyllingen 2012, 104–116). This indicates that various parts of the house had different functions, but is not sufficiently substantial to suggest that one part of the house was a byre. At Kvia, no such pattern is seen regarding the distribution of macrofossils and cleaned grains are distributed evenly throughout the house. This is interpreted by archaeobotanists Sara Westling and Anette Overland as evidence for human habitation only (2012, 33–34). The evidence from the two-aisled houses does, so far, not support the presence of a byre, although early experiments with ways of keeping animals inside the house cannot be ruled out.

The presence or absence of fireplaces is indicative of the internal use of space in three-aisled longhouses. Fireplaces represent the human living section, and activities associated with cooking and crafts. However, considering that the byre might have served different purposes at times of the year when animals did not live inside, fireplaces could also be found in the byre. Or, fireplaces might have been needed to keep fragile newborn lambs warm if they were born very early in the year and during bad weather.

The distribution of structures combined with the internal layout of space are tentatively suggestive traits in the architectural choices that point towards indoor keeping of animals in one part of the house, although this is not a traditional byre with stalls. However, strong substantiation of keeping animals indoors is the sheep coprolite found in a posthole in Austbø house 2. The evidence from this house allows for a different way of thinking regarding keeping animals indoors. I therefore suggest that the structural evidence embedded in the architectural choices is consistent with byres for the indoors keeping of sheep rather than with stalls for keeping cattle. Before developing this argument further I will study the broader use of the environment, as well as the faunal remains from Early Bronze Age Rogaland.

Figure 4.22 Map over shoreline levels in the Jæren region in the Early Bronze Age. Map by Theo Bell Gil / Museum of Archaeology, University of Stavanger.

Situating the Bronze Age farm in the wider environment

The architectural development from two-aisled to three-aisled longhouses in Rogaland coincides with changes in land use and also environmental changes (Prøsch-Danielsen and Simonsen 2000a, 2000b, Høgestøl and Prøsch-Danielsen 2006). Within a wider framework, a starting point to understand this transition is to consider the relationship between the two-aisled and three-aisled houses. In two of the cases where we see the transition at work, Kvåle and Kleppestemmen, a change in orientation signifies a break with tradition. Could this signify a new orientation in lifeways as well as in physical orientation?

The development in architecture cannot be separated from the environmental changes that happened in this period. In the Bronze Age, the sea level would have been 5–6 m higher than today and due to a warmer climate the area was covered with deciduous forest (see Figure 4.22). As agriculture was established, a process of deforestation started, which at the same time provided cleared areas for grazing and growing grains, and provided timber for the longhouses (Fyllingen and Armstrong 2012, 4–5).

Extensive deforestation is traceable in pollen diagrams (Høgestøl and Prøsch-Danielsen 2006), which may indicate that forest clearance took place to aid the grazing of domestic animals. The animals then upheld the initial clearance of the land and kept it open by their grazing. Otherwise, regrowth and reforestation would have happened quickly. With the management of animals becoming pervasive in the everyday lives of people, a greater toll on the environment is seen. In some places, unintended consequences of the intensified land use led to environmental changes such as deforestation, land erosion and the development of heathlands that still remain today (Prøsch-Danielsen and Simonsen 2000a, 2000b). Evidently, Bronze Age environmental changes were formed by a complex interplay between social, economic and ecological factors.

Environmental changes from the Neolithic to the Iron Age caused by human and animal agents by way of farming practices have been studied rigorously in south-western Norway, particularly in the Jæren region (Prøsch-Danielsen and Simonsen 2000a, 2000b; Høgestøl and Prøsch-Danielsen 2006; Lillehammer 2007; Soltvedt *et al.* 2007). Høgestøl and Prøsch-Danielsen (2006) observe two distinct environmental trends that are noticeable in all pollen diagrams from Jæren from the late Neolithic until the Early Bronze Age; deforestation and the development of pasture. A similar trend is seen in North-western Jutland (Kristiansen 1993, 29; Bech and Mikkelsen 1999, 69). These indicate an expansion in pastoralism and permanent fields. The environment was formed in a particular way by farming practices and the increasing importance of animal husbandry, which demonstrates that animal agents, by way of farming practices, were crucial to environmental changes (Mlekuž 2007; Armstrong Oma 2010).

The wider environment of Kvåle, one of the settlement case studies from Jæren, has been studied rigorously, by both archaeological and scientific methods. The two trends mentioned above, are clearly discernible in the period between 2000–1500 BCE: first, there is extensive deforestation of the original deciduous birch, elm and hazel forest by burning. This was visible as a preserved layer of charcoal in soil profiles. Dating of the charcoal demonstrated that this was a recurrent event that took place intermittently throughout this period. Additionally, pollen taken from preserved soil layers under clearance cairns show grasses and weeds that are consistent with grazing, particularly *Plantago Lanceolata, Rumex Acetosa, Asteraceae* and *Succisa* (Soltvedt *et al.* 2007, 151, 174, 198). Deforestation and intensified land use corresponds with Høgestøl and Prøsch-Danielsens (2006, 23, 32) "clearance phase 3," identified for the region of Jæren in the period 1900–1400 BCE, and interpreted as a direct reflection of the introduction, intensification of, and also preference for, the practicing of animal husbandry. Together, these factors demonstrate that the Bronze Age people altered the face of their land to facilitate, and better the conditions for, their domestic animals. And subsequently, the animals kept the landscape open by their grazing. Beyond economic considerations, this process has implications for a shared human–animal practice. By altering the landscape, humans and animals were, in a sense, domesticating the environment to their mutual benefit, thus deepening their joint life-world and reciprocal dependence on each other. Mutually embedded in a profoundly domesticated matrix, it seems highly likely that people could not imagine their lives without their animals, nor could the animals without humans (Armstrong Oma 2013c).

During the second phase, clearance of forest continued, and the altering of the landscape became further intensified. Stones were cleared into clearance cairns, creating smaller patches of more manageable fields. Pollen samples from buried land surfaces under clearance cairns show further increase in grasses and weeds such as *Cichorioideae, Campanula, Lychnis, Trifolium repens* and *pratense* and *Ranunculus acris* consistent with grazing and field mowing (Soltvedt *et al.* 2007, 113). Production of hay is consistent with indoor stalling of domestic animals, and a necessity for feeding the animals through the North-European winters. This could be in accord with seeds from weeds and meadow plants seen in the section interpreted as the byre in Tengesdal house 3. Worth noting is that in both phases there are no traces of fences or pens—seemingly the grazing herds roamed free in the landscape. Control over animals seems of less importance and perhaps bringing animals indoors was not merely a tool for herd management; rather, an ethics of care between humans and their animals resulted in a practice of seamless giving-and-taking that established humans and animals both as stakeholders in their joint environment (Armstrong Oma 2013c, 2018). Tangible implications of this are explored in chapter 5 and 6.

However, although the environmental data from pollen and macrofossils give clear evidence of significantly boosted numbers of ruminants in the landscape, these analyzes cannot establish species or breed. A mixed economy is often assumed, in which cattle, sheep, goats and pigs were kept.

Following the investigation of settlement sites and their environment, I presently turn to examine the evidence of domestic species present in the Early Bronze Age environment, before venturing into a closer examination of the conjectures to see what they can tell us about ways of living with animals.

Faunal remains from Forsandmoen and Kvåle

Due to the acidity of the soil, bone preservation is infamously bad in Rogaland. Therefore, very small numbers of fragmented bone are generally found. A small number of bones was recovered from Forsandmoen, altogether 306 from the whole life-span of the village, from the Early Bronze Age to the Migration period (400–600 CE) (for the report, see Hufthammer 2004). The bones could stem from any period of the site's habitation, and cannot serve to give an economic strategy for any one-time period. Also, they are not in a state to tell us much beyond species at best, or size equivalent to some species. The level of fragmentation suggests that they represent domestic waste, i.e. remains from meals. However, they give a strong indication of what kind of animals were consumed, and kept. Around 10% of the total number, and over half of the identified bones were most likely sheep/goat bones. Of the 61 identified bones, 28 were identified to sheep/goat, with a further 7 to small domesticates, which might also very likely be sheep/goat. 10 were identified to possible pig, due to their size, whereas small numbers of bird, fish, seal, beaver, deer and hare were found. Remarkably, no bones were identified to cattle, or large size domesticates. The presence of fish and seal, birds and wild animals like beaver, deer and hare suggests that resources other than strictly domestic were used. The site is located where the mountains meet the sea, and is perfectly situated for utilizing both marine sources as well as trekking up into the mountains for trapping and shooting game.

An even smaller number of bones was recovered from Kvåle, mostly these are human remains from what was thought to be the remnants of a grave. The bones were analyzed (Hufthammer 2002), but were mostly too fragmented to identify to species. Notably, a phalange from sheep/goat was found in a feature just east of House 3, the three-aisled longhouse at Kvåle (Hufthammer 2002; Soltvedt *et al.* 2007, 56). Admittedly, the assemblages are small, but it is still significant that there is an absence of cattle in both. I hold that this represents a real situation, and that sheep and/or goats were the main animals kept on these sites, as well as some pigs.

Zooarchaeologist Sean Denham took a look at the bones from Kvåle[8] and found a few fragments that could be sheep bones, but remarkably no cattle bones. Denham's experience as a zooarchaeologist led him to conclude that cattle was then not consumed, since their bones are big and dense and should be preserved better than sheep bones. The same conclusion is valid for the faunal remains from Forsandmoen. Therefore, at both the settlements of Forsandmoen and Kvåle, there are pointers towards the presence of sheep and the absence of cattle.

I will briefly mention the find of a probable sheep horn-core, found in a mound near a Middle Bronze Age house at Håbakken, Orstad, in Klepp.[9] The excavators expected the osteological remains in the mound to represent a human burial, but no human bones were identified amoung the bone fragments.[10]

To my knowledge, no dog bones are found in Early Bronze Age contexts in Rogaland. Since bone preservation is generally very poor, this is no surprise and does not in any way exclude dogs from Bronze Age life. Rather (see discussion in the following chapter), I suggest that dogs were not eaten, and thus their bones did not end up in the household waste, which is the primary source of faunal remains for archaeologists.

8. Personal communication Sean Denham 10.09.2010
9. Archived letter to Olle Hemdorff at Arkeologisk museum i Stavanger from osteologist Rolf W. Lie.
10. Personal communication Olle Hemdorff 01.10.2010.

The lands beyond: A glance at Skrivarhelleren

Skrivarhelleren is a rock shelter in the mountains of Årdal in Sogn, west coast Norway (see Figure 4.23) that was occupied from the Late Neolithic until Medieval times (Prescott 1991a, 1991b). The site was probably used as a part of a transhumance strategy, which is supported by the finds of bones from sea fish that have been brought up to altitudes of 790 m above sea, where the rock shelter is situated in the sub-alpine zone just below the timberline (Prescott 1991a, 37).

Figure 4.23 Skrivarhelleren and its wider context in southern Scandinavia. Map by Theo Bell Gil / Museum of Archaeology, University of Stavanger.

Table 4.2 Domestic species from Skrivarhelleren based on identified bone fragments (data compiled from Prescott 1991a, 40, tab. 1).

Mammals	LN/EBA	EBA	EBA/LBA
Canidae (dog/fox/wolf)	3		2
Sus scrofa (pig)			1
Bos taurus (cattle)	4	1	1
Ovis aries/Capra hircus (sheep/goat)	51	9	23
Total number of bones found	2175	591	1319

Five bone pins and one bronze fastener needle from a fibula are found at Skrivarhellaren, these are dated to the Late Neolithic and the Early Bronze Age. The pins are understood as indicative of the use of costumes made of woven wool textiles (Prescott and Melhem 2017, 318–319). Additionally, a number of objects are tentatively interpreted as tools for textile productions, such as bone plates with perforations, objects interpreted as weaver's shuttle or baton, bone needle with perforation and needle pins (Prescott and Melheim 2017, 319–321; Appendix). These finds are interpreted as indicative of both wearing of wool textiles and also textile production. On the basis of this, Prescott and Melheim suggest that wool production was an important trade commodity in trade networks and interactions with Southern Scandinavia in the Late Neolithic and the Early Bronze Age.

Providers of wool-sheep were also present at the site. An unusually large (in a Norwegian context) bone assemblage was unearthed (Prescott 1991a) and comprised butchery waste as well as many tools and implements crafted in bone (Prescott 1991b, 60–73). A wide variety of domestic, wild and marine species are present throughout the stratigraphic layers, which range from the Late Neolithic to Medieval times. The two species that are of interest to this book, dogs and sheep, are both present in the midden.

Table 4.2 presents the number of bones from identified domestic species from the stratigraphic layers dated to the Early Bronze Age and adjacent periods, and also gives the total number of bone fragments from the layers. The Bronze Age layers have yielded marked differently total numbers of bones, and the EBA layer has significantly lower numbers. All domestic species are included in the table to give an overall picture of the frequency of the different domesticates.

From the LN/EBA transition, three bones from dog (*canidae*) are found, and two from the EBA/LBA transition. Even though these numbers are low, they indicate the presence of dogs in the Bronze Age. Regarding sheep, there is a marked presence in all three layers. In the LN/EBA transition, a high number is found—51, and in comparison a much lower number for cattle, with only four. In the Early Bronze Age proper, sheep numbers drop, in tandem with the significantly lower general numbers from this layer, and then pick up a bit in the EBA/LBA layer. Cattle are represented with merely one bone in each of these layers. Pigs are represented by just one fragment alone from these layers. Similar to the bones from Forsandmoen discussed above, these numbers do not represent a post-depositional bias, rather, they give a realistic picture of the kinds of animal remains that ended up in this midden. Clearly, also at this site beyond the pale, high up in the mountains, sheep were by far the most important domestic species. Thus, we can infer that the situation on Jæren regarding faunal remains might be representative for the Early Bronze Age of Western Norway.

Goats?

A short note on the relationship between sheep and goats is necessary, since the faunal remains from Rogaland can not always distinguish between these two species, which have bones that are morphologically very similar. First, there is no direct evidence that places goats on the scene. On the contrary, where clear evidence positively confirms species, it points at sheep, such as finds of sheep droppings in the postholes of the house at Austbø, the rock art herding scene at Valhaug and the sheep horn core in a mound at Håbakken. However, this does not exclude the presence of goats. A useful comparison might be made with the site of Voldtofte on Funen in Denmark, in which a large and well-preserved faunal material was retrieved. Here, the relationship between sheep and goats is about one goat to every 10 sheep (Nyegaard 1996). Nyegaard suggests that sheep were probably exploited for milk, wool, hide and meat. A strainer was found amongst the pot sherds, used for making cheese, maybe from goat milk (Nyegaard 1996). Goats are also used as 'shepherds' in sheep flocks, as they are less nervous and generally have a positive influence on the flock (Nyegaard 1996, 40–45), Nyegaard (1996, 40–45) suggests that the goats could have been used as a social factor in animal flocks.

This might also have been the case in Rogaland. However, I suggest that sheep were the most important animals in terms of abundance, and the rationale underlying the transition from two-aisled to three-aisled houses was to facilitate a socio-economic contract between humans and sheep. This led to greater physical proximity between them, and facilitated greater control as well as greater intimacy. Clearly, larger flocks were sought after and pastoralism became more important.

And then there were sheep

Although there is little evidence as to which animals lived in the Bronze Age in the region, the evidence points unanimously towards sheep as the most important domestic breed. This does not exclude other domestic breeds, there is evidence for pigs and probably goats, and even though no cattle bones have been found, they might have been present.

On the basis of the presentation of the houses from Early Bronze Age period I and II, the environmental changes and the faunal material, and my discussion of the empirical data, I propose the following hypothesis: that *the transition from two-aisled to three-aisled longhouses is connected to, or even initiated by, sheep becoming household members.* This entails a closer relationship between humans and sheep, as well as a greater economic importance of sheep, larger flocks, and subsequently a need to make space for sheep inside the house.

I see this as a gradual transition, there are pointers in the Early Bronze Age two-aisled houses towards an internal separation of space, such as dividing walls, opposed entrances as well as small postholes in one end, like at Bjorhaug. The hybrid houses demonstrate that there is no abrupt break in tradition, rather the same group of people had different needs regarding their way of living and organizing their lifespace. Thus, they tested out new architectural solutions by creatively exploring and experimenting with ways of building, which reflects how needs can be expressed in architectural choices in a convenient way. With the three-aisled longhouse they found their formula, sufficiently successful to survive until the end of the Viking period, some 2700 years later (Armstrong Oma 2016c).

I will not exclude interpretations made by Horve (2009), that one part of the house might have been used for storage or other purposes and activities. Like I discussed above, the houses might have been multifunctioning and could serve other functions than byre during parts of the year when not occupied by animals.

In the following chapters, I will take a leap of faith and consider my hypothesis, that sheep became household members of the three-aisled longhouse—as correct. The chapters explore the consequences of this hypothesis, and consider why living with sheep might have been advantageous for both parts, and what its consequences were, in terms of human–animal relationships. The three-aisled longhouses were sites of repetitive physical encounters between humans and their animals, and the repetitive nature of the shared lifespace meant that the encounters solidified into particular relationships. For such relationships to form, openness must be present from the very beginning, as well as a willingness to expose oneself and the group to vulnerability and risk. At the cusp of the processes that led to the three-aisled longhouses, both humans and domestic animals were vulnerable in their openness to risk.

In order to grasp the consequences of my hypothesis, it is important to recap the position of sheep in Bronze Age Europe. Since management practices are vital to understand the life world that was created by human-sheep societies, I will discuss herding and explore the relationship between sheep and dogs by mapping the evidence for dogs in Bronze Age Scandinavia in the following chapter. Further, I will also discuss the changed sociality of sheep in connection with the so-called secondary products revolution.

— 5 —

A Closer Look at Sheep, Sheepdogs and the Dynamics of Herding

Beyond the "animal category"

After several years of working with theorizing human–animal relationships, I see a growing need in my own work to venture beyond the animal category and discover the actual animals behind it. This book is a response to that quest. In this chapter I go beyond the animal category to study the animals themselves, and their species specificities. A central tenet in my exploration is relationality; relationships not only between humans and animals, but further between different kinds of animals. However, looking to the past, the quest is fraught with difficulties since, like the people, all the animals have long ago passed on. Each bone fragment represents an individual animal, and several might come from the same animal, but each of these still represent the individual to whom they belonged. Nevertheless, I accept the limitations inherent in the material record, and on the basis of bone fragments alone it may not be possible to reconstruct the personhood, agency and ontological status of that individual animal. A study of the full context is key, and broad-spectral studies add invaluable information when attempting to understand the social and ontological status of animals. Still, it is important to fundamentally acknowledge the inherent sentient creature that the bone fragment once embodied.

The dynamics of herding

As outlined in the previous chapter, several factors in the archaeological record from Early Bronze Age Jæren point toward a substantial presence of sheep on the farm. The rock art scene of herding from Valhaug (Figure 1.1), as well as the presence of sheep bones, sheep coprolites and palaeobotanical evidence of intensified grazing, all point to a particular way of living in Bronze Age Jæren: sheep herding. This chapter discusses the presence of sheep and dogs in Northern Europe in the Bronze Age, both in terms of archaeological data and also considers historical accounts of traditional herding practices, in which shepherds, sheep and sheepdogs are bound together in the choreography of herding.

Additionally, I draw upon Human–Animal Studies research and the ethology of sheep and sheepdogs to deepen the understanding of what it means to live in close proximity with these two species, and what implications this has for an archaeological interpretation of the life on a farm in the distant past. I follow Louise Westling's (2014, 43) central tenet that in the practice of herding, all three species constantly "read" each other to be able to flow along together, and all three species are involved in a bodily communication that leads to cooperation (Westling 2014, 44). By cooperating, all three parties exert their agency.

From this standpoint, I discuss the social dynamics of herding, and how the intensification of herding strategies could have impacted life on the farm in Bronze Age Rogaland. Ultimately, the new husbandry strategies that are evident in the archaeological record are a step in the process of domestication, and brought humans and the domestic animals closer than previously. Understanding the processes of domestication of animals goes beyond simply taming and ensuing genetic changes in animals, but addresses new ways of living, cross-species, both politically and socially (Armstrong Oma 2018; Armstrong Oma and Birke 2013, 113; see also Mlekuž 2013).

The domestication of sheep

Sheep were domesticated in the so-called Fertile Crescent in the Middle East some 11,000 years ago according to Zeder (2008) and Colledge *et al.* (2005), and 9,000–10,000 years, according to Juliet Clutton-Brock (1987). The disparity in dating is due to new scientific methods that have dated early domestication events based on evidence for herd management and crop cultivation that predate the morphological changes traditionally used as evidence for domestication by at least 1000 years (Zeder 2008). Sheep were the first husbandry species to become domesticated (Ryder 1983). It is commonly assumed that domestic sheep (*Ovis Aries*) developed from the Asian Mouflon sheep (*Ovis orientalis*), and that the European Mouflon sheep are the feral offspring of an early wave of domesticated sheep that came out of the fertile crescent (Meadows *et al.* 2007). Regarding sheep, the *Ex Oriente Lux*-hypothesis holds true. Recent genetic studies into mitochondrial DNA (Hiendleder *et al.* 2002; Meadows *et al.* 2007) and endogenous retroviruses as genetic markers (Chessa *et al.* 2009) have demonstrated that there were several domestication events and that "sheep were recruited from wild populations multiple times" (Meadows *et al.* 2007, 1371). Also, studies of retroviruses have separated waves of sheep migrations into Europe, identifying "breeds previously recognized as 'primitive' such as Orkney, Soay and the Nordic short-tailed sheep now confined to the periphery of NW Europe" as relics of a first migration, whereas a later migratory episode involved sheep with improved production traits, which formed the basis for present-day breeds (Chessa *et al.* 2009, 533).

Although these studies reveal some clues as to how the domestication process took place, they say nothing of why. The reasons underlying domestication are complex, and the characteristics and needs of both humans and sheep must be considered. Fundamentally, sheep were originally socially well adapted to life with humans, with certain innate characteristics that were important fundamentals in human-sheep relationality: lack of aggression, manageable size, and a social nature (Budiansky 1992, 78–80).

On top of this, successful breeding was facilitated by early sexual maturity and a high reproduction rate. It is assumed that initially, humans kept sheep for meat and skin—although it is important to acknowledge that the reasons underlying the initiation of the domestication process are unknown and reasons beyond "meat on the hoof" could have been equally important. What has become evident is that sheep are one of the most successful species on the planet considering genetic development, diversity of breeds and geographical distribution. Starting from a small ecological niche in ancient Western Asia, they have conquered the world.

Following the initial domestication events, and the subsequent waves of migrations of domestic animals into Europe, sheep have cohabited with humans for millennia in Northern Europe. According to archaeological evidence, sheep were first introduced to Europe during the Neolithic (e.g. Marciniak 2005). In Norway, early indications of husbandry practices are reflected in pollen from *Plantago Lanceolata* and other plants that indicate pastures, the earliest ones are dated to around 4000 BCE (Myhre and Øye 2002, 27–31, with references). However, farming settlements do not appear until the late Neolithic, with the two-aisled longhouses. The earliest sheep remains from Western Norway are from the late Neolithic, dated to about 2500 BCE. They were discovered in the rock shelter known as Skipshelleren (Hufthammer 1994; Prescott 1996). Other early finds come from similar contexts, such as the rock shelter Skrivarhelleren, described in chapter 4. In Rogaland, sheep bones were found in the rock shelter Stangelandshelleren, and a cattle bone from the same context was dated to 3340–2890 BCE (Høgestøl and Prøsch-Danielsen 2006, 23)—although it is unknown if the sheep bones are from the same time. Caves and rock shelters provide good contexts for the preservation of bones, whereas settlements with two-aisled houses, when they start appearing in the late Neolithic, have very poor preservation due to the acidity of the soil. Therefore, based upon bones from rock shelters we can infer that sheep were present in Western Norway from the late Neolithic. Further, rock shelters, some of them far inland and high in the mountains, suggest the practice of transhumance from the Neolithic onwards (Prescott 1995).

Thus, already in the late Neolithic sheep were one amongst other domesticated species that cohabited with humans on settlements and formed herds that roamed the land, both the coastal regions where farms have been located and the mountainous inland regions where faunal remains in rock shelters attest to their presence. There is, as previously outlined, no evidence to suggest that the domestic animals lived inside the two-aisled houses, and loosely herded flocks that served as "meat on the hoof" in addition to hide for various purposes, is the most likely explanation of how sheep and other domesticates entered a social contract with the Neolithic people. However, we must bear in mind that the people and their flocks lived in tangled and complex relationships, and like Arkadiusz Marciniak points out, "Neolithic farmers could certainly not be viewed as 'dummy' economic people" (Marciniak 2005, 9). Still, assessing the full complexity of these relationships in Western Norway in the Neolithic, lacking the rich data sets from other times and other regions in Europe (such as Çatalhöyük, see for example Hodder 2014) is beyond the scope of this book.

Sheep and secondary products in Bronze Age Europe

Early Bronze Age Europe stood at the gates of innovations that would change the life of humans and animals: the *secondary products revolution*. The notion of a "secondary products revolution" (Chapman 1983; Greenfield 2005; Sherratt 1981) is still flourishing in the Bronze Age discourse. Sherratt (1981) first coined the phrase, and considered its economic and social repercussions, although his main focus was upon production from an economical perspective. The genetic studies mentioned above (Chessa *et al.* 2009) indicate that sheep with better wool production traits emerged in Europe at some point after the first introduction of sheep, and supports the "secondary products" hypothesis. However, these sheep were probably introduced to Northern Europe at a later point, considering that the Early Bronze Age wool textiles are made from fibres consistent with a sheep breed similar to Soay sheep (Ryder 1981; see also detailed discussion below), and Soay sheep are considered as relics of an early migration of sheep lacking the production traits introduced later.

Although the genetic studies referred to above are informative regarding some of the processes of sheep migrations, they cannot date migration waves. Archaeological evidence points to historically known seafarers in the Bronze Age, such as the Phoenician and other ethnic groups, as instrumental in spreading the sheep gene pool. Zeder (2008, 11597) suggests that "the expansion of domesticates and agricultural economies across the Mediterranean was accomplished by several waves of seafaring colonists who established coastal farming enclaves around the Mediterranean Basin."

However, the *secondary products discourse*—also adapted to discussions of genetic changes—entails that animals were passively regarded as food and products rather than as producers and agents, and the social practice implicit in the engagement with animals is rarely considered. My focus in the discussions in this chapter is centred around what the innovations had to say for people and animals on the cusp of a new way of being. This can to some extent be understood from looking at distribution patterns of species and age/sex in the faunal remains.

The age and sex ratio of flocks varies according to the different ends that one strives to achieve when managing a flock (Payne 1973), although the patterns observed in the faunal remains are debated (see Halstead 1998, 4–5 for discussion). For example, a flock of sheep intended for milking should consist of a few rams, and no wethers, and a large flock of rather small ewes, as more milk is produced from a large group of younger animals than a smaller group of older animals (Ryder 1983). If a meat production strategy is employed, animals would be raised to an economically viable size and no further. In comparison, pigs grow quickly and require less work in terms of producing fodder—pigs are, in this sense, more economic than sheep (Albarella and Serjeantson 2002; Serjeantson 1996, 2007), but they do not give secondary products. Although meat was probably initially the main product of the small ruminants, sheep and goats, innovations in farming led to an increasing importance of secondary products, milk, wool and hair.

It seems that at Forsandmoen secondary products held importance over meat, considering the significantly higher number of sheep/goat bones compared to pigs, the "walking larder." Sheep, on the other hand, were producers rather than products, a factor that could have given them the status of household members (Årlin 1999; Armstrong

Oma 2007). I envisage the enabling and facilitation of greater proximity, and thereby intimacy, between people and domestic animals as a result of the secondary products revolution.

The established presence of animals, in which tending them was part and parcel of utilizing both primary and secondary products (Chapman 1983; Greenfield 2005; Sherratt 1981), meant that animals became embedded in the society's habitus (Bourdieu 1977, 1990 [1980]). The secondary products were mainly material in their nature, such as milk, wool, hair, blood and manure (Serjeantson 2007). Traction is also considered a secondary product, although utilizing cattle and horses for their draught capacity can hardly be called a product. Likewise, horses, and possibly donkeys, were used for riding, which again is not as such a product but a particular set of actions. The change in habitus rests in how the materiality of the secondary products themselves, and the material culture surrounding the procurement, handling and storage of them, required new ways of engaging with the animals, and new sets of action in which both farmer and animal needed to partake. Therefore, I understand the full situation surrounding secondary products as *sets of action*. Such sets of action had material and immaterial consequences: they established new kinds of practices that both humans and animals engaged in and subsequently a material culture that facilitated them, as well as the (immaterial) idea that animals could be producers. Secondary products were not utilized consciously before the Early Bronze Age. Therefore, these sets of actions constitute a re-negotiation of the social contract between humans and domestic animals at this time (Armstrong Oma 2010), a development which required both humans and animals to explore new possibilities within their shared practices and find how to best accommodate each other within these new expectations.

The use of these sets of actions led to a new materiality, a range of artefacts such as ceramic vessels, milk strainers and draught equipment for equids and cattle (Halstead 1998; Sherratt 1981), and I argue, new architectural solutions. In northern Europe, Serjeantson (2007) has demonstrated how such sets of actions became increasingly important in Middle and Late Bronze Age Britain, as the percentage of sheep and pigs increased during this period. The same trend is discernible in Southern Scandinavia (Kristiansen 1998, 108). Changes in settlements and an increasing intensification in agriculture, resulting in permanent fields, are interpreted as outcomes of intensification and specialization in the management of animal husbandry in Middle- and Late Bronze Age Britain (Serjeantson 2007).

In most parts of Bronze Age Europe, sheep/goats were the third most important animal, whereas in the coastal regions along the North Sea and the Baltic Sea, the Alpine regions and partly in the Carpathian basin, sheep/goats ranked second, after cattle (Benecke 1994, 133). In Denmark, they were generally the second most important animals, although sometimes pigs have been found in larger quantities (Nyegaard 1996).

Little data may be found regarding the age/sex ratio of sheep/goats in Bronze Age Europe but, where information is available, the bones indicate adult animals, suggesting that secondary products—milk and wool—were important (Harding 2000, 254). Finds of products made from wool in the coastal areas of Middle and Northern Europe imply that wool was utilized (Benecke 1994, 133), and presumably given particular importance. However, zooarchaeologist Georg Nyegaard, who examined the faunal remains from Bronze Age Denmark, did not find conclusive evidence for extensive use of secondary products based on sex/age ratio of sheep bones (Nyegaard 1996, 151–152, 160–161). But his conclusions are mainly based upon the large Late Bronze Age assemblages, since faunal remains from Danish Early Bronze Age sites are, like for Rogaland, scanty. Faunal remains from sheep were present in smaller quantities at the three Early Bronze Age sites included in his study (Torslev, Bjerre and Rumohrsgård). At Bjerre, where the largest assemblage was found, sheep remains were excavated from contexts dated to Early Bronze Age period II. Sheep teeth from younger individuals were found, and also two teeth belonging to adult individuals, one of them at least 3½ years old (Nyegaard 1996, 12). These finds suggest an economy based in part on secondary products. Still, Nyegaard's results point towards an economic strategy dominated by cattle, since faunal remains from cattle dominate the assemblages, in particular Bjerre. This is consistent with finds of three-aisled houses with traces of stalls at this settlement (Bech 1991; Nyegaard 1996, 13). This demonstrates that Rogaland is out of step not only with the geographically close Danish Bronze Age societies, but also with the general trends in Europe, since sheep appear to be the most important animal in Rogaland at this time. The importance of sheep indicates that their products were highly valued.

Sheep wool was a valuable commodity in the Bronze Age. Recent studies of the body and costume of the so-called Egtved woman from the Early Bronze Age (Frei *et al.* 2015) have given new insights into the wool trade. Studies of isotopes of separate wool fibres from the woman's wool garments and blanket demonstrate that the fibres belong to sheep that come from, and have been grazing in, different geographical locations in Europe. This means that the textiles were made from a composite of wool from sheep from different places, and thus that wool was exchanged over large distances (Frei *et al.* 2015 supplementary information, 18–19). Recent analyses of the wool textiles from several individuals in present day Denmark that have been dated to the Bronze Age demonstrate that the Egtved costume is not a singular instance. Karin Frei and her team (Frei *et al.* 2017a) analysed textiles from 10 burials. The analyses showed that mostly, the textiles were created from wool with isotope signatures that do not match Danish baseline signatures, but came from sheep that had been living in other parts of Europe. Further, the isotope signatures vary, which means that the wool might not have one origin but rather come from several places in Europe. Wool trade must therefore have been more common in the Bronze Age than previously assumed. These results strengthen Klavs Randsborg's (2011) suggestion that textiles were a major trade item in the Nordic Bronze Age and that wool formed part of long-distance European exchange networks.

Rather than being a homespun product, where the local flock would provide wool for the clothes of the community, textiles were, like bronze, composites of materials from widely different geographical regions. Materiality from sheep was thus the raw material for commodities that were considered very valuable and, as such, lent something beyond its function as clothes to their wearers. Frei *et al.* (2015, 2017a) do not suggest which regions the wool might have originated from. Wool from Rogaland might have formed part of the Egtved girl's—or of any of the other well preserved bodies'—garments.

Obtaining wool from so-called primitive sheep requires particular sets of action. Primitive sheep, like Soay sheep, shed their fleece and it is possible to either pluck (roo) it by hand or to wait for it to shed naturally and pluck it from shrubs and undergrowth (Figure 5.1).

Figure 5.1 Hand rooing in prehistory. This illustration succinctly exemplifies how much trust was needed on behalf of the sheep to allow the kind of handling that sets of actions for procuring wool requires. Drawing by Flemming Bau. Rendered with permission from the Museum of Archaeology, University of Stavanger.

However, with the latter method a quantity of wool will be lost to the pasture and to birds' nests, and the quality of the wool is considerably lessened since the fibres are disturbed by dirt, twigs, leaves, and so on. To get a fleece that is relatively clean and even, with fibres that are smooth and even, the best way is to hand pluck the fleece, known as rooing. Primitive sheep breeds cannot be shorn, since their fleece contains a fibre called kemp, longer than the softer wool of the fleece. Advantages to rooing are that the wool growth from the previous year is removed at the right point in time, which leaves next year's growth to rise cleanly on the sheep, and there is no loss in length of fibres.[1] Rooing can be done by inserting spread fingers into the fleece and then literally lifting the fleece off, provided the sheep sheds evenly. When working progressively along the sheep, it is possible to lift off the fleece more or less intact. Loose tufts in the fleece normally indicates that the sheep is ready for rooing, and it is advantageous to stay close to the sheep to know when time is nigh. Being able to roo a sheep by hand requires a strong bond and mutual trust, the sheep must cooperate by staying still, letting themselves be handled and allowing the humans to do the work. Studies of fragments of Bronze Age textiles from Rogaland (from Blodheia on Karmøy and Jåsund in Sola) demonstrate that the tips of the fibres were present, which is only seen in cases where the wool has been rooed, as shearing removes the tips (Rast-Eicher and Jörgensen 2013, 1236–1237, tab. 6).

Genetic studies like the ones cited above suggest that in all likelihood the Bronze Age sheep in Rogaland were remnants of one or several of the early migration waves, and their modern counterparts can be found among northern European "breeds previ-

1. See http://www.okacres.com/2013/05/some-sheep-are-so-rooed.html and https://www.youtube.com/watch?v=B8hp-rngcEA

Figure 5.2 Soay sheep of several wool colours at the Cranborne Ancient Technology Centre. Photograph by Simon Barnes, CC BY-SA 2.0, https://commons.wikimedia.org/w/index.php?curid=3499513.

ously recognized as 'primitive'" (Chessa *et al.* 2009, 533). Based on a different study of mitochondrial DNA, archaeologists Antoinette Rast-Eicher and Lise Bender Jörgensen (2013, 1240) suggest that "Scandinavian Bronze Age sheep are linked to a branch coming through Siberia to the north of Europe." This is based on the presence of a group linked to Asian heritage found in the Volga region, the Nordic countries and in Russian Karelia (Tapio *et al.* 2006). Unfortunately, it has proved impossible to extract ancient DNA from Scandinavian wool textiles (Rast-Eicher and Jörgensen 2013, 1240).

Wool fibres from Scandinavian Bronze Age textiles come from hairy sheep, a type of sheep which have a fleece composed of short fibres of fine underwool, without medium fibres, and an outer coat of coarse hair with large kemp, with a yearly moulting (Rast-Eicher and Jörgensen 2013, 1225, 1227). Based on wool fibre analysis, Michael Ryder deemed the contemporary Soay sheep to be the closest living remnant to Bronze Age breeds, a feral breed that live on the island of St. Kilda, off the coast of Scotland (Ryder 1981, 382). They are hardy and can live off pasture that is considered marginal. Modern breeders[2] still recommend that they are cared for and offered shelters or housing. The wool is rooed (hand plucked) rather than shorn. Morphologically, Soay sheep appear to be similar to the flock depicted on the rock art panel on Valhaug in Rogaland (see Figure 1.1), with lithe bodies and slightly curved horns, although with shorter tails. The appearance of both Soay sheep and the flock on the rock carving is very different from the modern archetypal sheep—a bundle of wool.

Soay sheep are one of the breeds identified by Chessa *et al.* (2009) as relics of the early waves of migrations, prior to the later sheep with better production traits (Figure 5.2).

2. http://www.oldlinesoays.com/basiccare/

Thus, if Early Bronze Age sheep were similar to Soay sheep, they were on the cusp of a process that resulted in breeding that led to a bettering of the genetic production traits regarding secondary products.

The current diversity of sheep breeds demonstrate that, throughout the time between the Neolithic and the present day, sheep have undergone changes both regarding morphology and fleece, and additionally regarding skills and temperament. We can infer that Bronze Age sheep were different from most of today's common breeds both regarding morphology and temperament. They were neither docile nor very wooly, and probably much fitter, stronger and more aggressive in the sense of being able to fend for themselves. They were very likely also more inquisitive than most sheep today.

A closer look at sheep: Insights from ethology

Sheep are in modern culture frequently grossly underestimated regarding their cognitive capacities and abilities (e.g. Pryor 2006 [1998], 99)—they are considered to be totally witless, following their flock blindly. But research has proven such prejudices to be flawed. Pryor (2006 [1998], 99) puts it thus: "Contrary to popular belief, sheep are far from stupid; in fact much of their so-called 'stupid'—seemingly irrational—behaviour may closely be linked to their strong herding instinct." The flock is actually a powerful social unit in itself, in which individuals bond strongly. Flocking behaviour is not a mindless activity, rather it is the best possible adaptation to face predators, since sheep have no means of defence except escape and safety in numbers. Social flocking behaviour that acts both as an anti-predator and grazing strategy are learned social traditions (Festa-Bianchet 1991). Also, as pointed out by Hatkoff (2009, 124) "the bonds sheep form amongst themselves help them manage the environment and cope with stressful events."

Following Wemelsfelder, Hunter, Mendl and Lawrence (2001), sheep are, like other animals, "behavers" and can be assessed as such. To describe behaviour as an expressive process, the following example is given:

> a ewe separated from her lamb in the hills may walk about with her ears pricked up, looking out and bleating loudly, all the while appearing "agitated," observed expressive state, that is as a state apparently reflecting the animal's experience of the situation it finds itself in. The animal's expressive state may change: when the ewe finds her lamb, she will become calm and more relaxed. Thus qualitative assessments can capture fluctuations in behavioural expression that are otherwise hard to record... (Wemelsfelder *et al.* 2001, 210)

This demonstrates that a close observation of sheep can facilitate an understanding of what they express and further what they experience. Such a doorway to understanding was probably used by people in the Bronze Age, who lived so close to them.

Recent research has shown that there is more to sheep than meets the eye. There are amazing similarities in how the sheep brain and the human brain process visual information (e.g. Hatkoff 2009, 124–136). Sheep are capable of forming strong social bonds, and recognize individuals of both their own and other species (Estevez *et al.* 2007; Fisher and Matthews 2001)—even after long periods of absence. Visual communication is vital for flocking behaviour, sheep use their ability to recognize the other members of the herd to monitor herd behaviour and make choices according to the movements of the others, thus staying together as a flock. In the words of Dimitrij Mlekuz (2013,

152): "Anyone watching a flock of sheep moving across landscape, changing shape but always staying together as one, knows how mesmerizing it looks." Pryor (2006 [1998], 96) has observed that sheep become more docile and easier to manage when they are closely confined in a group, and herd animals become distressed when they are left alone; "remove the herd, and you remove their protection, or, more important, their sense of security." This instinct has great time-depth, and Pryor (2006 [1998], 99) holds that it is a basic urge also for wild or feral sheep to gather together tightly when they feel threatened from the outside.

Sheep can learn the lay of the land; the boundaries, safe and unsafe zones, and understand safety and danger as being conditional to different times of the year. Such sheep are known as "hefted sheep" by English sheep farmers since they know and understand the "heft"—the grazing ground (Law and Mol 2010; Rebanks 2016). This knowledge is something that sheep as a flock take years to build, and is then transmitted from ewes to lambs in generations, as a socially transmitted memory. The "heft" is as old as the flock, which means it can go back thousands of years. Etymologically, "heft" comes from an old Norse word, "hevð." This word means tradition, and also rights to land or property based upon tradition. In spite of its Norse roots, sheep farmers in Norway do not use the term heft today, although they have firsthand knowledge of what the concept entails.[3]

In Norway, where transhumance is still practiced, sheep are intimately acquainted with the wild mountainous areas where they are left to graze during the summer months, and the need to know and understand the "heft" is a trait that presumably dates back to their wild predecessors (Børresen 2004, 156). Innumerable anecdotes exist regarding the uncanny ability of sheep to find their way home. The importance of the knowledge of the "heft" was starkly highlighted during the foot-and-mouth disease in Britain, when whole flocks of sheep were annihilated and this knowledge was lost (Law and Mol 2010, 71).

Brain research has demonstrated that sheep have a remarkable memory and can remember the faces of 50 other sheep for over two years and "the specialized neural circuits involved maintain selective encoding of individual sheep and human faces even after long periods of separation" (Kendrick *et al.* 2001). Distinguishing faces is a social task, one that we obviously share with sheep. Being able to distinguish faces "afford us a remarkable ability to discriminate between, remember and think about many hundreds of different individuals" (Kendrick *et al.* 2001, 165)—so also for sheep. Kendrick *et al.* suggest that sheep and humans alike have the ability to remember and respond emotionally to individuals in their absence. Further, they respond positively to seeing pictures of smiling and calm human faces, and react with stress to angry and upset faces—even when these are associated with food rewards. They also calm down when stressed when they see pictures of familiar sheep. They also distinguish between human faces that were "morphed" by a computer to appear very similar, and can detect subtle differences in faces and facial expression, just like humans can (Kendrick 1998). Despite the common prejudices against sheep intelligence, they have remarkable abilities that research is constantly revealing.

3. Personal communication Tarald Oma 20.11.2016.

On the basis of this research, Mlekuž (2013) makes a convincing case arguing for complex societies among sheep. Sheep perform their society through their bodies, and the tools available to them are, in addition to their intelligence and history, their social skills and strategies. These are "soft tools" and cannot, according to Mlekuž, fix and stabilize society. Their societies do not have durable institutions but depend upon the social transference of knowledge and tradition from one generation to the next, as was demonstrated by the foot-and-mouth disease in Britain. Nonetheless, by use of their social skills, sheep live in a society with their own kin. Sometimes their society necessarily includes other species, too, and in Bronze Age Rogaland, humans and dogs were members of sheep society.

Dogs in the Bronze Age

Dogs were one amongst other species that were a part of sheep society in the past as well as in the present. The rock art from Valhaug, shown in Figure 1.1, demonstrates that the practice of herding holds great time-depth. Dogs in the Bronze Age were mostly medium to large—thought to have been either court dogs, watch dogs, or hunting dogs (Benecke 1994, 133)—and, I would add, herding dogs. In Scandinavia, dogs have been found in domestic deposits in, for the main part, Late Bronze Age houses and pits in settlements, and rarely show any sign of having been butchered or eaten (Nyegaard 1996 Appendix, 49–53; Ullén 1996, 153). A close relationship between dogs and humans is often presumed (Ullén 1996, 160), but it is left unclear what this postulated proximity implied. The modern myth of the dog as Man's best friend is often, rather simplistically, directly transposed upon the societies of prehistory.

In the Bronze Age, there is less consistent evidence for dogs in the archaeological record compared with sheep. Evidently, dogs were treated differently to sheep in death. There could be several reasons for this, such as food taboos and/or that dogs held an ontological status as companion species, and thus their bodies were treated with respect.

To date, the most informative site regarding the status of Bronze Age dogs in Scandinavia is Apalle in Sweden (Ullén 1994, 1995a, 1995b, 1996). Due to fortunate circumstances, this settlement site was very well preserved, and unusually high numbers of bones were found in situ. Of all the bones, 1.3% were identified to dog, and these were of a spitz breed (Ullén 1996, 150). Dog bones were found at the site from both the Early and Late Bronze Age phases, and in both periods, dogs were clearly non-food, as there are no butchery or cooking marks on their bones (Ullén 1996, 153).

A particular depositional pattern is seen in the Early Bronze Age phase of the settlement: Dog bones were deposited with care in pits with fire cracked stones at the periphery of the site termed the "outermost boundary of settlement" by the excavator, Inga Ullén (1996, 150). In addition to these special deposits—"dog pits"—scattered elements of dog bones were also found spread all over.

In the Late Bronze Age phase, (Ullén 1996, 151), such "dog pits" were found near houses, often one near each house, either by the end wall or near the entrance. There was always a cranium in the pit, giving them a "grave-like character, with dogs being deposited individually or in pairs" (Ullén 1996, 151). Ullén (1996, 153) also reports of houses in which dog jaws were buried in pits underneath the clay floor or in postholes either side of the entrance of the house (Ullén 1994, 1996; 1997, 67). Based upon this depositional pattern, a close relationship between humans and dogs is presumed (Ullén 1994, 1996, 1997).

An examination of faunal remains from Bronze Age Denmark shows that only small bone assemblages have been retrieved from Early Bronze Age sites (Nyegaard 1996, 9–14). For example, only a single dog tooth was retrieved from Torslev. At Bjerre, the identified cattle bones have gnawing marks from dogs, and thus attest to the presence of dogs. However, there are some sizeable Late Bronze Age sites with large bone assemblages, most notably the middens from Kirkebjerget, Voldtofte, where high numbers of dog bones are present (Nyegaard 1996 Appendix, 49–54). The assemblage contains the largest number of dog bones found in Scandinavian Late Bronze Age contexts. Mainly the dogs were deposited separately in pits, and not amongst other bones in the cultural layers. Most of the bones are intact. The dogs were of varying age, from young to old dogs. Their sizes indicate that they were about the size of a modern German shepherd, or even bigger. Some cut marks indicate that some dogs had their head cut off or that the carcass was butchered. Presence of dog is also indicated by gnawing marks on 5.8% of the bones of other species. In more than half of the cases, bones with gnawing marks are secondary butchery waste (where the meat was first consumed by humans). Primary butchery waste (head and extremities, parts with little meat) is less frequently gnawed. Nyegaard (1996 Appendix, 49–54) suggests that the dogs were probably working animals, assisting shepherds, since their large size would be suitable for shepherding large flocks.

I have not come across faunal remains that have been identified as dogs in Bronze Age Rogaland. However, as the faunal remains are so scant and seem to be remains of meals, this does not suggest that dogs were not present. But perhaps it points to a dietary taboo—maybe dogs were not eaten. This, at least, is the conclusion drawn by Gräslund (2002, 167), since she notes that dog (and horse) bones are hardly ever found at settlement sites, but are abundant in wetland sacrifices. For example, at the sacrificial site of Östra Vemmerlöv in Scania, dogs were beyond comparison represented with the highest number of individuals (van Post 1919). Ullén (1996, 146) states that altogether 20 graves with dogs are found in Bronze Age Sweden, mainly per II–III. This attests to the special role dogs held in Bronze Age society.

There is some evidence that materiality from dogs was used to ornament human bodies. An example of this is a drilled dogtooth that was found in the burial of Löderup 15, Scania, probably used as a pendant or part of the costume. The grave is dated to Montelius per. II (Bergerbrant 2007, 122, see also fig 89, p. 110).

Figure 5.3 Dogs from Flote, 1, with numbers according to the panel. Illustration by Kjersti Vevatne (1996, 26).

The rock art panel with a herding scene from Valhaug on Jæren is portrayed in chapter 1 (see Figure 1.1). The motif is unique and nothing like it is found on Scandinavian rock art. There are other portrayals of dogs and humans, but none to match the exceptional clarity and distinctive rendering of this scene. We see a little herd of four horned animals, all of them moving as one. Behind is a shepherd with arms outstretched that drives the herd forwards. Coming up wide behind is a dog that crouches low, with tail held low. The dog and the right, outstretched, arm of the shepherd form a barrier against straying sheep, and together shepherd and dog work as one.

Dogs are found on other rock art sites in Western Norway besides Valhaug, particularly further north along the coast. Two examples are Flote 1, Etne, and also Bakke, Jondal. At Bakke, one possible dog is seen (Wrigglesworth 2011, 74–75). At Flote 1, four carvings which appear to be dogs are seen (see Figure 5.3 and Figure 5.4), all of them have paws, three are depicted with upstanding tails and one has a hanging tail (Vevatne 1996, 26–27). The dogs with tails that stand up look just like today's Norwegian herding breed Buhund, whose tails curl up.

In addition to the few rock art panels in Western Norway that portray dogs, more are found in Sweden, often these are hunting scenes (e.g. Bertilsson 1994, 95, 118; Fredsjö 1969, 115, 117, 128). They range from vague canine shapes to more specific renderings that give indications of breed. One such is the rock carving from Fossum in Tanum, that indicates

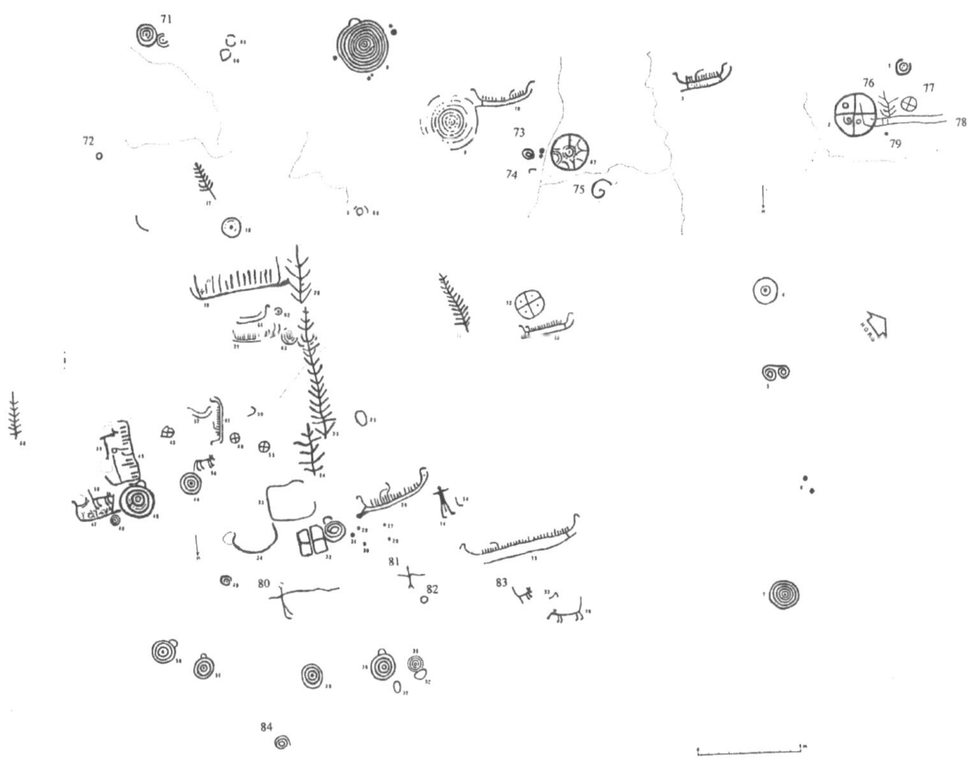

Figure 5.4 The panel of Flote 1. Note that two of the dogs are close to a ship, one is inside a ship and one is close by a sun symbol. Illustration by Kjersti Vevatne (1996, pl. 6).

Figure 5.5 The bronze dog (?) from Ree in Time, Jæren. Photograph by Terje Tveit / Museum of Archaeology, University of Stavanger.

breed: with "erect ears, curled tails and steep angulation of both forequarters and hindquarters they are similar to modern spitz dogs" (Gräslund 2002, 167; fig 2, 169). This is true also for the dogs at Flote. The astute stance of the Norwegian herding dog breed Buhund, with tail curled up and ears upright is also seen on a rock carving from Hamn in Bohuslän (Fredsjö 1969, 117, Bild 43). Here, two dogs stand between two archers, men and dogs face each other in a pose that could be interpreted as hostile. There are also other rock carvings where dogs are portrayed. In contrast, the dog at Valhaug keeps his tail low, adding to the impression of the sheepdog crouch with the low tail a sign of intense concentration. A rock carving from Rickeby in Uppland has depictions of at least two dogs in a similar pose (Ullén 1996) indicating that the rock carvers paid acute attention to the dogs, both as regards breed, personality and situation.

In addition to the close observation and intimate knowledge of dogs as individual animals that rock art motifs demonstrate, there are broader implications. In the archaeological literature, rock art is perceived as loaded with mythological significance (e.g. Goldhahn 2006, Goldhahn *et al.* 2010). Depicting dogs—and moreover a herding scene—in the cosmological sphere of rock carvings merges lived human-animal relationships and the choreography of herding with the realm of the otherworld.

Finally, worth mentioning is that a bronze figurine from Ree in Time, Rogaland, in my opinion depicts a dog, see Figure 5.5. Due to lack of clear morphological traits, this animal figurine has been construed as both a horse and a calf (Kristoffersen and Armstrong Oma 2008), however I find it more likely that the artist intended to portray a dog, due to the short neck and head held low. If my interpretation is correct, this figurine adds momentum to the importance of dogs in Bronze Age Rogaland. Bronze was, at the time, an extremely precious metal that was traded from far away places that must, to Bronze Age people in Rogaland, appeared as beyond the pale. Mary Helms' (1988) ideas of the inalienable value added by long journeys into mythical time and mythical space resonates in this context. Casting the figure of a dog in bronze thus attests to the high value and respect for dogs in society.

A closer look at sheepdogs

We cannot know for certain what types of dogs were about in the Bronze Age in Rogaland, since the faunal remains do not give any clues. However, by way of the rock art at Valhaug, and other places in Norway and Sweden, we know, as demonstrated above, that dogs were present in society. And judging by the Valhaug motif, some dogs were sheepdogs, and are even portrayed working in a way that bears an uncanny likeness to the working style of modern sheepdogs—Border Collies.

Contrary to popular belief (referred to by Pryor 2006 [1998], 98), the Valhaug rock art demonstrates that prehistoric sheep were indeed worked by dogs. Pryor (2006 [1998], 98–100) holds that in his experience, so-called primitive breeds such as mouflon sheep can easily be worked with dogs, and there is no reason to suggest that prehistoric sheep were not herded with the aid of dogs (Figure 5.6). The sheep need to be, in his words, "dogged" meaning that the sheep will learn to move together and form a flock when a dog approaches (see also Westling 2014, 49). An ethnographic study from Rogaland of sheep and transhumance to mountain pasture quotes shepherds that claim the same thing: sheep need to be familiar with dogs in order to respond to herding as a choreography, otherwise they will scatter to all the furthest and most inaccessible corners of the mountain and chaos ensues (Grimstvedt *et al.* 1996, 196–197).

This process is a habituation and socialization process between sheep and dog in which their natural instincts—for sheep to flock, and for dogs to hunt—are moulded

Figure 5.6 This photograph demonstrates that soay sheep can be herded by dogs. Photograph by Priscilla Weaver / Saltmarsh Ranch.

Figure 5.7 Border Collies are famous for their supreme herding skills. Photograph by Anne Helen Robberstad / Jærmuseet.

to create herding. It requires more moulding and training for the dog than it does for sheep. Therefore, an in-depth discussion of the making of a sheepdog is appropriate, since this would have been a major challenge for the Bronze Age shepherd.

To investigate the complexity of the phenomenon of sheepdogs, I will give a comprehensive presentation of the modern-day Border Collie, and explore how this breed is specially trained and bred to use their herding instincts to work sheep. Although there is no known genetic or morphological link between Bronze Age dogs and modern dogs, herding—the work at hand—is fundamentally the same, and thus similar chacacteristics in the dogs were sought after. I choose an in-depth discussion of the Border Collie rather than the Buhund due to their working style, which is reminiscent of the Valhaug motif. But I do not thereby imply that today's Border Collies are related to Bronze Age dogs—merely that an in-depth study of Border Collies and their working style can shed light on past herding practices. I follow Pryor, who holds that even though modern sheepdogs are a "new" breed, there is nothing "new" about what they do: "Collies have been bred to accentuate this gathering-up or herding behaviour, but the point I want to make is that it's not anything new or unusual. It's not a trick invented by some cunning medieval monk or circus animal trainer. Prehistoric farmers could perfectly well have observed the same behaviour in their own dogs themselves. In fact I'm convinced that's precisely what they did" (Pryor 2006 [1998], 100). I suggest that the Bronze Age farmers in Rogaland were equally perceptive regarding the skills and abilities of their herding dogs as their contemporaries in the English Fenlands that Pryor studies.

Today, the Border Collie is the sheepdog *par excellence*, and is used for shepherding all over the world (Figure 5.7). The breed has been popularized by competitive sheepdog trials, and TV programmes such as BBC's "One man and his dog"[4] but most of all by the remarkable intelligence and skills found in most individuals of the breed. Some extraordinary traits set the Border Collie aside from other working sheepdogs: the silent use of

4. A somewhat misleading title, since many women are active participants in sheepdog trialling.

"eye" to control the sheep, the crouching creep and the wide-circling outrun to gather the flock.[5] It is so compelling and breathtaking to watch these dogs do their job that trialling has even been turned into a spectator sport. The Border Collie is frequently acknowledged as the most intelligent dog breed (Coren 1995), and in Animal Studies, several Border Collies have become famous for their remarkable aptness to quick learning and problem-solving. For example, they can have vocabularies of hundreds of words, and learn to associate new words with objects or tasks after just one repetition. Therefore, research on "smart animals" has to some extent focused on Border Collies. One of these is Chaser, who has a documented vocabulary of 1022 words (Pilley and Reid 2011). In instances like those, the Border Collie enthusiasts might legitimately claim that any other breed is just a dog.

In addition to their remarkable intelligence, they are also extremely energetic, acrobatic, athletic and agile. These qualities make the breed highly suitable for sheep herding, as intelligence, agility and physical stamina are required for long days of work, often in difficult weather and challenging terrain. The shepherd might be too far away to see, or to give signals, and the dog must solve problems by her/himself, such as getting sheep off dangerous cliff ledges (Kennard 2004, 2005; Longton and Sykes 1997; Scrimgeour 2002; Sykes 1999).

Border Collies have been bred for their working skills rather than according to a breed standard of physical appearance, which has secured the favourable traits that make them so suitable for herding. However, the breed cannot claim pedigree back to the Roman Empire, in fact the breed was registered as late as 1915. Nonetheless, dogs that look like Border Collies are found on works of art that go far back in time.[6] As sheep herding has great time depth, going back to the secondary products revolution and beyond, so probably do sheep dogs have great time depth. The name of a breed is less interesting that the qualities of the breed, and what kinds of tasks can be accomplished with those qualities. The Border Collie is interesting because in these dogs, such qualities are not only extremely well developed but also so well documented. Further, the rock art panel from Valhaug (Figure 1.1) shows a dog in a pose that is strongly reminiscent of how modern day Border Collies work: the pose is low—crouching, almost, and the dog is coming around the flock at the back. This is typical herding behaviour, and this is how these dogs work, they creep along the ground, behind the sheep, weaving back and forth to gather up any stragglers. They must be firm, and often use their eyes face-to-face with sheep in staring contests. The Border Collie eye is a phenomenon by itself, these dogs will stare you down and use their unflinching gaze to bend the sheep to their will. Archaeologist and sheep farmer Francis Pryor puts it thus in his blog: "The key attribute any sheepdog must possess is a good strong eye. Basically this means that the dog's stare affects sheep and makes them uneasy. They can't relax."[7]

Another interesting factor regarding Border Collies is their heritage and recent history. The breed was, as mentioned above, recognized as late as 1915, however the first

5. http://www.bordercolliemuseum.org/EarlyHistory/EarlyHistory.html
6. See for instance "Twins" by Sir Edwin Landseer (1802–1873). The title refers to the two lambs. Of the dogs, one is sable and one is tri-colour. http://www.bordercolliemuseum.org
7. http://pryorfrancis.wordpress.com/2012/02/09/on-sheep-and-working-dogs/

Figure 5.8 Old Hemp. http://www.bordercolliemuseum.org/AuldHemp/AULDHEMP.html

sheepdog trial was earlier, in 1873. The name might stem from Coalie, which might mean black dog,[8] or from a Celtic word for useful.[9] Border refers to the borderlands between England and Scotland. But probably the breed has been around for a long time before its official recognition, as farmdogs and herding dogs. The Border Collie Museum[10] has an online collection of evidence showing dogs that look like modern Border Collies on depicitions that go back to the 1500s.

In the last 100 years, probably at least in part because of the popularity of sheepdog trialling, the traits so characteristic for the working Border Collie, the strength of eye, the crouch and the wide outrun, have been nourished to create some very talented dogs. Most of today's Border Collies can trace their ancestry to Old Hemp (1893–1901), a trialling champion that exhibited the characteristic herding traits and went on to sire many litters (see Figure 5.8). Allegedly, his handler, Adam Telfer, said of him: "He flashed like a meteor across the sheepdog horizon. There never was such an outstanding personality." His working style became the sought-after standard of Border Collie herding (Halsall 1985). These dogs work the sheep in a particular way, that is beneficial to the sheep flock, as is aptly described by Sheila Grew:

> characterized by its ability to move large or small numbers of sheep in a silent controlled manner in complete co-operation with its master... [and] a most useful asset possessed by many of these border working collies was the power of the "eye," the ability to control

8. http://www.bordercollies.nl/eborhis.shtml
9. http://en.wikipedia.org/wiki/Border_Collie
10. http://www.bordercolliemuseum.org

> the sheep by staring at them in a fixed and steady manner. Dogs with the right amount of "eye" can keep their sheep bunched together well when driving them and thus avoid a great deal of flanking, running from one side to the other. This in turn keeps the sheep calmer and so they are less fatigued. (1993, 1)

What I find interesting about the history of the Border Collie, and that serves well as a potential parallel for sheepdogs in the Bronze Age, is that the qualities that today are prevalent have been nurtured consistently for not much more than a century. The development of Border Collies as a breed owes much to the dedication and commitment of the shepherds who cared enough to breed only the best. For without them and their knowledge the Border Collie as we know it today would not exist.[11]

A similar scenario can be imagined for the Bronze Age: the intensification of keeping sheep and the need to herd larger flocks, in a changing landscape, led to seeking out new qualities in the dogs that worked the sheep. Those could be a gentle manner, a strong eye, a wide outrun—ways of working that allow the sheep to deal with the stress of herding. Above all else, the ability to work in close cooperation with the shepherd would be vital, and the rock art from Valhaug attests to a close working relationship. And as the need arose, with the onset of intensified sheep husbandry, such qualities could be nurtured and be brought to blossom within a few generations of dogs. Thus, it could be that dogs such as Old Hemp have forebears in ancient times in the deep history of shepherding, flashing meteors that we will never know.

The Norwegian equivalent to Border Collies is the Norwegian herding dog breed Buhund, which literally means "dog in the dwelling." The Buhund is a spitz type, and is an all-round farm dog used for herding, and as watch-dog (Thorsen 2001, 98–106). Archaeological evidence has established that dogs that morphologically were very similar to Buhund were present in Scandinavia in the Iron Age (e.g. Lia 2001), demonstrating that the breed has deep roots.[12] The Buhund has a different working style compared to Border Collies, they have a stronger chase-instinct and drive the flock away from the shepherd rather than rounding up. Additionally, they bark—they do not work silently. These characteristics are suitable for gathering up large flocks from mountain pasture, but less so for closer interactions on the farm (Grimstvedt *et al.* 1996, 196–197).

Historical accounts of keeping sheep in Western Norway

Rogaland is still one of the regions of Norway in which sheep are held in the greatest numbers (Grimstvedt *et al.* 1996). The landscape lends itself to sheep rearing, with its short distance between flat lowlands and mountainous areas with rich summer grazing.

In his magnus opus, *Sheep and Man*, Michael Ryder (1983) wrote in detail about historical sheep rearing in the west country of Norway, and his observations might also very well be true for prehistoric times. He says that: "the main husbandry in Norway has long been one of winter housing and mountain transhumance from the western coastal and valley regions, with flocks of breeding ewes from twenty to fifty head kept on mixed farms" (Ryder 1983, 528–529). Ryder's description of the year of the sheep is in accord with Kennard's (2004, 2005). The year starts in November, when the sheep return from

11. http://www.fostbc.org.uk/about-the-breed-1.asp
12. http://www.buhund.org/Breed/View/History

Figure 5.9 Transhumance is still widely practiced in the Norwegian mountains. Photograph by Anne Helen Robberstad / Jærmuseet.

summer pasture in the mountains, and mating takes place at the end of this month. Over the winter the sheep are brought into sheep-sheds at night, but left to graze all over the farm during the day. "In the past older sheep grazed heather all winter with no extra food and no shelter" (Ryder 1983, 529). Haymaking took place at the mountain cottage [*seter*] and was transported down to the farm on a sled during winter. December to March was a tough period, where foodstuffs such as branches, bark from logs, and in coastal areas seaweed and fish offals were fed to sheep to tide them over. Then in April, lambing would take place, when the grass was beginning to grow again.This was still a difficult month because the ewes needed extra food since they were either in late pregnancy or needed to provide milk for their young. From May onwards the sheep would enjoy extensive grazing during the day, but ewes and lambs were still housed at night. The sheep were shorn between March and May, normally the latter. In June, large flocks were gathered together under the care of one shepherd and taken to pasture in the mountains, in a fixed transhumance pattern. They were driven up by foot, a journey that could take a week. Shepherds would stay with the flock at summer pastures, but still a loss of 2% was considered normal. The shepherd would use sheepdogs such as Border Collies or Buhund (Norwegian sheepdog breed) to work the sheep. At the end of summer the sheep were gathered with the aid of dogs and driven back to the farm. The lambs were weaned when they return to the farm, and butchery took place in October.

The narrative presented by Ryder (1983, 528–529) is partly framed by the yearly cycle and climate of Western Norway, aspects of life that have not changed much since the Bronze Age. Some parts of the narrative are confirmed by the archaeological data. One of these is the fixed pattern of transhumance (Figure 5.9). Finds of Bronze Age temporary settlements in rock shelters in the mountains, such as Skrivarhelleren, demonstrates that pastoral transhumance was most likely also practiced in the Bronze Age (see discussion in chapter 4, and Prescott 1991b, 1995, 1999). Second, is the use of dogs when herding sheep as testified to by the Valhaug rock art panel (Figure 1.1). Third, I argue, is the housing of sheep and thus sharing roof with humans at particular parts of the year, particularly during lambing and for some time after until the lambs were better able to fend for themselves. I suggest that the three-aisled houses in Rogaland were built for this particular purpose. This will be explored in depth later in this chapter and in the final chapter.

Ryder's narrative combines an approach that is economically viable and simultaneously centred around care, and in some periods of the year a proximity and intimacy between sheep and humans grows from the intensity of the husbandry practice. Such periods are lambing, shearing, butchering and mating. It is also clear that to the degree that sheep were housed indoors, this was not a question of control, but rather to facilitate care and protection, from both predators and harsh weather. Ultimately, the care was for the sake the health of the young, the life force of the flock and thereby the future of the herd. In the periods when less care was needed, greater freedom was given to the sheep. Interestingly, Ryder outlines a mutually constituted working relationship, based upon trust and care. The outline of a social contract is seen, in which care and protection is combined with an economically viable husbandry strategy to uphold the life-force of the flock.

Considering Law and Mol's analysis of the Cumbrian sheep (discussed in chapter 3), the flock is something that is bigger than the sum of its parts. Flocks of sheep are the result of long-lasting relations of care. Law and Mol state that there is pride in the history of breeding, which is the result of the careful selection and care that go into raising a flock. One of the daily practices of farming is to take care of the flock (2010, 64). Thus, the flock of sheep is invaluable, outside of value (2010, 66). Also in Bronze Age Rogaland, care was most likely the most important daily practice.

I understand such practices under the umbrella-term *ethics of care* (Armstrong Oma 2018). This term is associated with *duty of care* and is normally used in social sciences and particularly in medicine and nursing to denote the duty and practices of protecting, nurturing and caring for those that are weak, sick, injured or disabled. But is also a much wider term, and in UK legislation, duty of care is implemented in the Animal Welfare Act.[13] In a broader sense, it denotes the duty of behaving in such a way to others as to not do them harm, but to protect them—and it is in this sense that I extend this notion to animals in the past.

Particular human-sheep practices emerged from a way of living, and in the Bronze Age, one of these was that humans and sheep shared their living space, and the implicit care invested in this spatial arrangement. A deep mutual respect potentially grew from such an interaction, and even though the outcome of the relationship for the sheep was their slaughter, this was presumably treated as an act of sacrifice (see Mlekuž 2013). A closer look at the sheep-human relationship discloses practices that go beyond a simplistic domination regime. The domination dogma that has been spun around human-domestic animal relationships (see Ingold 2000a) must be unbound and looked

13. https://www.legislation.gov.uk/ukpga/2006/45/section/9

at with a fresh perspective. Rather than one species controlling and dominating others, extending agency to animals (see chapter 3) demonstrates that there is subtle dynamics at work in human-dog-sheep relationalities.

The triad of shepherd, dog and sheep: The advantage of early socialization and indoors rearing

Louise Westling is informed by semiotics in her discussion of the zoosemiotics of herding sheep with dogs, and how to train herding dogs (2014). She underlines how important it is that the dogs

> *agree* to cooperate in this game or dance, and the trainer must earn that agreement [...] The human trainer must "read" the dog's behaviour, just as the dog must attend to and "read" the trainer's behaviour. (Westling 2014, 44, orginal emphasis)

She leans on Deleuze and Guattari (2004 [1987], 262–263) and their "becoming-animal" in which humans become animals and animals become humans "by moving in a block of synergy and thinking *with* the other" (Westling 2014, 44, orginal emphasis). She goes on to say that adding sheep to the situation of training dogs brings great complexity into the block of becoming, because the three species are all involved in bodily communication that leads to cooperation.

The dynamics between sheep and dog is a performed choreography in which both parties know and act their part. It can be described thus: A flock of sheep is grazing in the distance. The shepherd—known as the handler, by professionals, signals the dog/s to perform the *outrun*,[14] and the dog scoots off at full speed, running around and behind the sheep, giving them wide berth as to not frighten and scatter them. Then begins the *lift*, the dog/s gather the sheep by weaving around and behind them to take advantage of the natural flocking instinct of sheep (Festa-Bianchet 1991) and then *drive* them in the direction the shepherd commands, normally "away to me,"[15] i.e. right towards the handler. Often obstacles and gates must be traversed on a drive. When the dog and sheep get to the handler, the drive can continue, or dog and handler could together *pen* the sheep, or *shed* (separate) one or a group of sheep by *singling* them out from the flock. The sheep flock draw upon their sociality, they choose to cooperate and allow themselves to be driven, thereby exercising their agency. The most important factor for success is not the skills and training of the dog, but the mutual socialization of dogs and sheep. If the sheep are not familiar with dogs, they will not let themselves be lifted, driven or penned by a dog (Grimstvedt *et al.* 1996). This brief description serves as a demonstration of how herding consists of particular sets of actions that were prerequisites for the practice surrounding the procurement of secondary products.

Socialization is a cornerstone for the herding practice, and requires particular spatial arrangements that allow humans, sheep and dogs to get familiar with each other. One of the aims of this study is to figure out what happened when people in the Early Bronze Age began to build three-aisled houses, and understand the rationale behind this break in architectural tradition. I argue that socialization is key to understand-

14. Terms in italics are terms that are in normal use amongst professional sheepdog handlers and in trialling.

15. Sheepdogs need to know left (often the command "away") from right ("come by").

ing the reasoning behind the transition. As demonstrated in the previous chapter, it is likely that the three-aisled houses accommodated both humans and animals. Here, I draw upon Human–Animal Studies to suggest the reasoning behind such a choice.

Keeping animals inside the house offers some clear advantages: notably, the animals become habituated with people, which allows for a development of mutual trust between human and animal on an intimate level that would never happen with flocks kept outside. With dogs, it has frequently been demonstrated that socialization and habituation to humans from birth and in the first period of life is crucial to develop positive feelings towards, and the ability to trust, humans. A British study demonstrated that 68% of dogs that frequently showed behaviour dominated by fear-aggression had been reared outdoors in a stable, in a yard or a shed, etc, and *not* inside houses together with humans (Appleby *et al.* 2002; Bradshaw 2011; Solbakken 2010). Amongst dog trainers it is well known that this early socialization, in dogs from birth until 16 weeks, is key to develop the kind of healthy social security that is so vital to create a strong working partnership (e.g. Bradshaw 2011; Price 2011; Westling 2014, 46). Some canine behaviour specialists urge dog buyers to only aquire puppies that have been reared inside the breeder's home (e.g. Price 2009, 23). As I have argued previously, Bronze Age farmers kept dogs, and so they must have known of the positive effect of rearing the young close to the hearth. Sheep farmers that breed sheepdogs today also rear the puppies indoors for the first few months of their lives.[16]

On the basis of this, I argue that indoor stalling of animals, especially of the young, has a profound impact upon the animal's ability to develop positive feelings and trust towards people, thus establishing a good working relationship, that makes it easier to handle the animals later in life. Good stockmanship demands the development of a positive relationship between oneself and the animals in one's charge, this makes sense both regarding animal welfare (Dawkins 1980) and gives higher production yields (e.g. Seabrook 1972, 2001).

Research on early socialization in different animal species has shown that there is a window of opportunity, and its time-length varies from species to species. Scott suggests that lambs "form social attachments to their caretakers and associates" within the first 10 days of life (1962, 16), whereas Markowitz *et al.* (1998) finds that in lambs, such attachments can be formed possibly only in the first three days (Figure 5.10). The same goes for cattle, a study by Krohn, Jago and Boivin (2001) indicate that handling and hand-feeding, especially during the first four days after birth, increases the motivation of young calves to approach a human compared with calves that receive minimal human contact.

Despite artificial selection for tameness (Price 1984), a tendency to maintain some flight-distance and physiological stress responses to people persist in domestic animals (Grandin 1987). For example, routine handling procedures of sheep, such as shearing and herding, leads to elevated stress levels (Markowitz *et al.* 1998, 573). Hutson (1985) has demonstrated that in sheep, stress can be alleviated by gentle handling and food rewards. Although tameness is to a certain degree heritable, environmental influences are important in animal's responses to humans, and one way of obtaining greater tame-

16. Personal communication David Kennard August 2011.

Figure 5.10 This lamb has accepted a human as a secondary caregiver. Photograph by Anne Helen Robberstad / Jærmuseet.

ness is through increased contact and socialization with humans in early life (Kilgour 1987).

Studies have been conducted to assess whether early human handling of lambs leads to increased affinity for humans in lambs. Markowitz, Dally, Gursky and Price found that "handling and feeding of lambs, especially in the first three days of life, can strongly influence their responses to people as they mature" (1998, 580–581). Their study also shows that lambs can, in this early period, form a primary attachment to their mother, and a secondary social attachment to human handlers, and that the secondary attachment does not interfere with the primary attachment (1998, 584–584).

Considering that lambs have a window of opportunity for socializing with other species and forming secondary social attachments with them that possibly only lasts the first three days of their life, it makes sense that Bronze Age farmers might have brought heavily pregnant ewes into the house to take advantage of this opportunity (Figure 5.10). It would be a gamble to leave the ewes outside, since it could be very difficult to get mother and lambs indoors within this time frame. The three-aisled house was well suited to accommodate ewes and newborn lambs, since the greater amounts of roof-bearing posts would have made it easy to make simple makeshift pens, and move groups of mothers and lambs around, much like sheep farmers do today (Figures 5.11

Figure 5.11 A meeting of lambs, from one pen to the other. Photograph by Anne Helen Robberstad / Jærmuseet.

and 5.12). The one section with more densely spaced trestles, that is found in many of the houses, would be very suitable for keeping sheep indoors in a flexible manner, since the denser posts would accommodate easy making and unmaking of pens. Likewise, the smaller postholes that are scattered inside the houses could be traces of poles that were moved around to make and unmake pens.

Further, socialization as a reason behind the transition in architectural choice also supports the transitions to three-aisled houses with stalls that are found elsewhere in Northern Europe and are presumed to accommodate cattle (e.g. Barker 1999; Bech and Mikkelsen 1999; Fokkens 1999; Rasmussen 1999; Zimmermann 1999b) since socialization with humans already from birth is also highly advantageous for calves (Krohn *et al.* 2001).

Figure 5.12 Contemporary sheep farms in Norway have a flexible system for makeshift pens where new families are kept during and after birth. Photograph by Anne Helen Robberstad / Jærmuseet.

Bronze Age herding—new ways of living

In historical and modern societies, our language reveals that domestic animals are at the heart of dwelling at the farm—consider the word **hus**bandry, in which hus- stems from house[17]—and likewise sheep are the corner stone of the herding practice. We can extrapolate from the words **shep**herd and **sheep**dog that the fact that sheep lend their name to the other kinds of beings in this triad demonstrates that sheep is the significant part. The outcome of the choreography is the secondary, and in the end primary, products from sheep. Sheep as producers are the valuable assets that the shepherd and the sheepdogs protect. As such, the vitality and well-being of the flock rests with the sheep (see also Law and Mol 2010, 64). The socialization between the three kinds of beings is, I argue, key to understanding how particular sets of actions, and further practices, developed—both sharing roof and herding.

The herding scenario demonstrates that an animal is not just an animal, rather, different animals are different *kinds of being*, and fulfill roles that are specific to them. In this particular scenario, sheep are "animals" shepherds are "human" and dogs are somewhere in between (Figure 5.13). The job performed by dogs in herding is of such importance to the shepherd that for most sheep farmers, dogs are indispensable.

17. Likewise, in Norwegian, the word for domestic animals is *husdyr*, which literally means house animals.

Figure 5.13 The sheepdog weaves back and forth between sheep and shepherd, in a complex choreography that bind them all together. Photograph by Kristin Armstrong Oma.

Sheep farmers will often describe the dogs as family members[18] that are allowed to live out their life on the farm, and are well cared for in their retirement. In his guide to training sheepdogs, Derek Scrimgeour (2002) talks of the importance of developing a "special relationship" between dog and handler, one that is based on trust and mutual understanding. His method of training is grounded in understanding the dog's point of view and setting aside the human interpretation of a course of events. Rather, when starting a young pup it is the job of the handler to help the dog understand the situation and what is required of her/him (Figure 5.14). But, says Scrimgeour, "The biggest influence, however, will be how the sheep react. All your dog wants to do is to control the sheep so it is up to you to help him get control" (2002, 26). The kinds of scenarios described by Scrimgeour and other handlers illustrate the slippery nature of such a three-way relationship, and demonstrate how greatly the shepherd depends upon the dog to create successful herding practices.

Westling (2014, 48–49) is more specific than Scrimgeour (although she refers to him as the expert), and points out two particular factors that must be present for successfully training a sheepdog: authority and mutual respect. She accentuates that a basic starting point is to work with the dog as a partner, and elements such as sureness, consistency and fairness are important. Trying to dominate the dog does not cut the mustard, and respect is, as she says, "necessary on all sides" (2014, 48), including of and from the sheep.

Sheepherding holds a timeless quality—the dilemmas of training, and the joys and frustrations of guiding dogs and sheep into a working relationship have been around since humans first started herding sheep and goats, some 11,000–10,000 years ago in The Middle East. As this way of living transformed lives of Bronze Age farmers in the

18. See for example the youtube movie "Dogs that changed the world (Border Collie)." http://www.youtube.com/watch?v=Ii1WnS_qa7g&list=UUsyzekLPZER0UDWYpfkWUqg&index=4

Figure 5.14 Starting up a young dog requires patience, skill, and innate qualities in the dog. This young dog is interested in sheep, and tests the strength of her eye, before tentatively setting off after the sheep, in a crouch with the tail kept low. Photographs by Kristin Sinnes.

hilly Southwestern Norway, dogs would have been absolutely indispensable to gather up the sheep and bring them safely home.

Dog are so vital to control sheep, and in the hilly terrains of Western Norway one might wonder if it is at all possible to herd sheep without dogs—the same is also true for the hilly regions of the fertile crescent where it is believed that sheep were first domesticated. Would it have been possible to embark upon the process of domestication without the help of dogs? Dogs hold such an important place in the history of the human society, since they fearlessly perform tasks that their human companions could not do by themselves. Some of the essential roles that dogs have played through history are as hunters to feed us, guards to protect us, and herders to ensure the lifeforce of flocks of other animals. Other roles have been given less attention but were just as vital—such as healers, spiritual companions and guides into uncharted lands. Today, the idea and implementation of ownership over dogs is embedded in our society, not only in language, but also in the eyes of the law. However, I wonder if guardian (see Irvine 2004) is not a better term to use regarding human-dog relationships in prehistory. Rather than being an owner, the term guardian acknowledges that humans are custodians and caretakers with responsibilities to the animals they share their lives with, but the animals are not their property (Irvine 2004, 64–65).

We see that animals on the farm existed together in complex social relationships, they were aware of each other as social partners of different kinds and with different kinds of status. What status did these three kinds of beings have? The number of individual sheep made up the flock, the life force, the producers of the precious wool and milk, and probably also served as a source of meat, considering the sheep/goat bones that were the remains of meals from Forsandmoen. Dogs probably played many roles, ranging from companions, hunters, herders; in various ways watching over the other kinds to keep them safe. They were a buffer between predators (for example wolves, their own ancestors) and prey—the sheep. The humans were the strategists, the masterminds—bringing the different kinds of beings together in a particular way to strive for the desired outcome. Providing for all, they tried to create a choreography that created a lifeworld in which these kinds of being were intertwined and dependent upon each other.

Westling (2014, 50) suggests that "herding of domestic sheep by dogs and their human partners is a special case in which more coordinated kinds of cross-species co-operation and communication have become habitual and have blurred species boundaries." Considering the Bronze Age, co-habitation should be added as a factor that was instrumental to this meshwork of beings, in which the in-between lines of communication and actions were as important as the agents. A perfect description is Haraway's "knot of species co-shaping one another in layers of reciprocating complexity all the way down. Response and respect are possible only in those knots, with actual animals and people looking back at each other, sticky with all their muddled histories" (2008, 42 and Figure 5.15). I envisage such knots forming in Bronze Age Rogaland, both in the three-aisled longhouses and in modifications of the landscapes, like the fields that were cleared to accommodate grazing.

Figure 5.15 Nimble feet and an agile mind is needed to herd sheep in this kind of landscape. Photograph by Anne Helen Robberstad / Jærmuseet.

Bronze Age ontologies: Sheepdogs, sheep, shepherds, and their status of being

A study of the participants in the choreography of herding raises questions of an ontological nature, in particular as concerns the ontological status of the participants. Introducing several kinds of animal agents and giving them unique, separate ontological status, challenges the way academia commonly labels species, as humans—us, and animals—others. Breaking down these labels raises questions of a philosophical nature, questions that allows archaeologists to open the doors of perception and get new insights into the rationale of past societies. Following Alberti (in Alberti *et al.* 2011) and the discussion of relational ontology in chapter 3, I hold that studying particular past societies from an ontological perspective gives us the opportunity to insert a difference in the way we understand past and present human-animal relationships.

The famous spiritual teacher Echardt Tolle talks of dogs as spiritual guardians of being, and in an interview[19] he talks of how humans experience a true, spiritual joy in the (random) meeting with a dog. I find the concept "spiritual guardian of being" interesting to think with regarding the nature of dogs and the roles they play in human lives. Numerous studies have shown that people who live with dogs (and cats) have better chances of recovery after traumatic illness, and have better overall mental and physical health, and in addition report that they feel a higher degree of contentment compared with people who do not live with dogs (see Allen *et al.* 2007 for a list of referenced stud-

19. https://www.youtube.com/watch?v=_C6CnSNdTuk

ies). These studies are conducted in current societies, and one might question if such "feel-good" qualities have any relevance for the archaeologist that studies the past. One explanation for the "feel-good" vibes associated with dogs is in our shared biochemistry, as outlined in chapter 2. When humans and dogs interact, a hormone called oxytocin is released both in their and our bloodstream (Handlin 2010; Nagasawa *et al.* 2009; Olmert 2009). This hormone is in popular science dubbed the "love hormone." The biochemistry of our forebears, and also of the ancestors of modern dogs, has been unchanged for millennia. The shared biochemistry accounts for the possibility of mutual socializations. Nevertheless, important to bear in mind is that a shared biochemistry does not explain the specific shape that such relationships take.

Within the ontology of herding, there are three kinds of being, each with their unique ontological status. The human shepherd acts as the self-proclaimed agent, the flock of sheep are by category animal others, whereas the sheepdogs are in-between. The latter can simultaneously be both necessary tools and "brothers" to the shepherd. Sheepdogs are therefore beings on the fringe, both categorically and in practice. They are in-between the human and animal categories, and as such, they contest a sharp separation between human and non-human animals. A key understanding of the ontological status of dogs as an "in-between" species, not human and not quite animal, can be found in Human–Animal Studies research. The biologist John Bradshaw proposes, in his book *In Defence of Dogs*, that dogs are unique in the sense that they can bond very intimately with humans, and at the same time retain their "dogness," their species specificities, for example regarding choice of mating partners (Bradshaw 2011). They do not become imprinted by humans in a biological sense, but are able to retain different kinds of relationships to different species, and distinguish between what kind of relationships are appropriate for which person, taking species into account.

The work that sheepdogs do often happens in dangerous, hilly "fringe" landscapes, weaving up and down steep mountain sides (Figure 5.15 and Figure 5.16; see also the link in fn. 17). Likewise, their role as agents in-between shepherd and sheep is at the ontological fringe, since they fall outside of clear-cut human versus animal categories. Studying their fringe position brings a different dynamics into human–animal relationships that are often tacitly understood to be an act of relating between two kinds of beings. What is seen here is a complex interplay between three kinds of beings, that mutually constitute their life-world. The position set out here is that there can be, and often indeed are, more than two kinds of being in human–animal dynamics.

When more than one kind of agent work together, a key aspect to their success is early socialization and habituation to each other as different kinds of beings (Grimstvedt *et al.* 1996; Pryor 2006 [1998]; Scrimgeour 2002; Westling 2014). The types of agents need to grow into being together. The three-aisled longhouse provided the spatial setting in which such a process could be facilitated—where Haraway's (2008) knots of beings could co-shape one another.

By all accounts, dogs had been in a mutually constituted relationship with human partners for thousands of years when this transition took place. The sheep were the new kids on the block as household members. But also the dogs had to find a new way of being in the world—they needed to master the transition from being dogs to becoming sheepdogs. In sheepdogs, the original hunting instincts have been harnessed and

Figure 5.16 In this kind of terrain, the sheepdog will often be far away from the handler and acts independently to herd rather than hunt the sheep. Photograph by Anne Helen Robberstad / Jærmuseet.

moulded to herd and protect. In the herding situation, sheepdogs are their own agents, they engage in a social world and are responsible for performing tasks from the situation of individual, autonomous problem solving—yet without making the kill. Indeed, the shepherds draw upon these capacities that are inherent in good sheepdogs when they train the dogs. A good sheepdog handler will socialize the dog to be a free agent (e.g. Westling 2014), and even though commands and communication are important factors in training, the most important factor is the development of trust and intuitive understanding between humans, dog and sheep.

Ontological status of being in the Bronze Age

I will pull out two phenomena from the archaeological record to examine the ontological status of dogs in Bronze Age Scandinavia. First, I wish to comment upon the depositional patterns of dog bones unearthed at the settlement of Apalle. Second, I will return to the rock art at Valhaug (Figure 1.1), in the heart of my case study area.

In the Early Bronze Age phase of Apalle, dog bones were deposited in pits around the edges of the settlement, and seemed to mark the boundaries of the settlement (Ullén 1996, 150). In the Late Bronze Age phase, dog jaws were in some instances buried by the entrance to the house, and otherwise in special dog pits, one near each house (Ullén 1996, 153). None of the dog bones show signs of butchery marks or other marks to indicate that they have been eaten (Ullén 1996). The treatment of dead dogs at Apalle demonstrates that their ontological status set them apart from the other animals, since their remains were not mixed with other species. However, there is a change in the depositional practice from the pits that demarcate the boundaries of the settlement to pits near the houses or by the entrance to the house—although both situations serve to separate the inside from the outside. This suggests to me that the dogs were communally protecting the settlement, and as such had the role of guardians of the community. Their ontological status is thus doubly as beings on the fringe, they hold the physical fringe between inside and outside, and they are also neither "animals" (i.e. not eaten nor deposited together with animal bones from other species) nor humans. In the later phase, they retain their ontological status as guardians, but seemingly no longer of the community at large. Rather, their proximity to individual households suggest intimate human-dog relationships. The presence of dogs within individual households is exemplified by dog jaws buried by the entrance posts in some of the houses. Their position in this scenario is possibly less on the fringe, and closer to humans. However these changes are conceived, it is clear that in both the early and the late phase, the ontological status of dogs is neither human nor "animal." They simply defy the normative categories created by normative science.

Rock art is an important source of understanding past social dynamics, as it presents past perceptions of not only the kinds of activities that people engaged in and found of sufficient importance to portray, but also shows how the artists, by their habitus, *saw* the kinds of beings that were present in their world. The herding scene at Valhaug suggests to me that the artist saw three kinds of beings, and depicted them very clearly. Ommund Møllerop, the archaeologist who first described the scene, was adamant that the artist deliberately set out to depict three kinds of beings, the four horned animals as one group, the human, and then the dog (Møllerop 1968, see also Nordenborg Myhre 2004, 94–95). This demonstrates to me that the artist, who reflected the general, local perception of the life-world and its rationale, *saw*—and thus experienced—these beings as three kinds. Subtle signals can be read from the positioning of the bodies: on the one hand, the shepherd's extended arms almost touch the dog as the dog is coming towards the human. Together, they form a closely knitted team. On the other hand, the flock of sheep is moving as one, away from the human-dog team. The flock forms another tightly knitted team. The fact that they move as one suggests that they have been, in the words of Francis Prior, "dogged" i.e. are familiar with the choreography of herding. The scene can be construed in two ways, either human and dog forming one kind of status against the "animal" status of the flock, or the three species forming individual ontological statuses. However, drawing upon evidence of the treatment of dogs and humans in death, the lack of Bronze Age dog graves suggests that dogs were not ontologically construed as humans. Thus, the scene suggests a three-way dynamics with three kinds of ontological status, the flock, the guardian (dog) and the shepherd.

Before I discuss the significance of my results in the next chapter, I want to dwell for a moment on the ontological status of sheep. Like I have discussed earlier, Mol and Law (2010) talk of what they call the "farming sheep" as having value only as a member of the flock, whereas "no individual sheep matters very much, but as a flock they are of immense and irredeemable value"[20] (2010, 65). Nonetheless, they appreciate the single sheep's agency (e.g. discussion in chapter 3). I agree that the life-force of the flock is immensely important, and the toil and care that went into altering the Bronze Age landscape in Rogaland to facilitate grazing for the flock testifies to this. As does the building of the longhouses, in which I believe sheep were (to larger or smaller degree) living. However, I wonder if Law and Mol's dismissal of the individual sheep would be applicable to this Bronze Age society, and I doubt it. The co-habitation of the longhouse might very well facilitate one-to-one relationships between humans and sheep. Consider bottle-fed lambs that are brought into the house—they become intensely socialized and form deeply trusting relationships with their caregivers. Such bonds might have been one of the unintended consequences that happened following the transition from two-aisled to three-aisled longhouses—differences made by agents of other kinds are not predictable. Thus, the ontological status of sheep was possibly not simply as a flock—a mass of blurred faces—but additionally as individual beings.

The following chapter elaborates upon the consequences of this kind of lifeworld for the Bronze Age people of Rogaland—the sheep people.

20. Worth noting is that Law and Mol are here discussing sheep in England during the foot and mouth disease, during which entire flocks were destroyed.

— 6 —

The Sheep People: Towards An Archaeology of Ontology

"Animals came from over the horizon"

The intention of this book is to grasp a diversity of perceptions of ontological status, rather than to establish *the* ontological differences. In previous chapters I have outlined an argument that presents my theoretical standpoint and how I see such statuses as simultaneously being in-flux and fixed in particular time-space. When humans relate to animals they have never been indifferent to them, but have construed their ontological status according to their cosmological world-view and underlying rationale. Some even propose that humans have a need to be close to animals, and the natural world—this is known as biophilia (for discussion, see Irvine 2004, 29–32). In his essay "Why look at animals," John Berger beautifully describes the fascination animals have held for humans through the ages:

> Animals came from over the horizon. They belonged *here* and *there*. Likewise they were mortal and immortal. An animal's blood flowed like human blood, but its species was undying and each lion was Lion, each ox was Ox. This—maybe the first existential dualism—was reflected in treatment of animals. They were subjected *and* worshipped, bred *and* sacrificed. (2009 [1977], 15–16, original emphasis)

In coming over the horizon, animals transgress the unseen separation, and enter into the life-world of humans. Berger presents animals in this context as though they possess an inherent dualism from the point of view of human perception. This can, of course, be a possible point of entry into understanding human–animal relationships. However, I find complexity to be a far better approach than duality, when coming to grips with the emergent becomings that spring from lived human–animal relationality, and the entangled knots (Haraway 2008, 42) that ensue.

As has become evident during the previous chapters, the archaeological evidence overwhelmingly proposes that the species that more than any came over the horizon and came to belong in Early Bronze Age Rogaland, were sheep. There is also evidence to suggest that sheep overstepped the boundary into the life-space of humans, and became co-habitants inside longhouses.

In this chapter, the evidence put forward in the previous chapters is assessed together, and I outline the social contract (see Amstrong Oma 2010) in which sheep, dogs and humans participated: the contract of herding. Further, I get closer to the women and men who created the society of the sheep people, by investigating their graves and where possible, their biographies.

But first, I want to dwell upon the notion of animals that come over the horizon, and use this way of looking at the emergence of intensive sheep farming to frame this chapter. In chapter 4, it was established that sheep were present in the Neolithic not only in Rogaland, but in Western Norway, and sheep were therefore known to the Early Bronze Age farmers. Questions that arise from a perspective of sheep appearing over the horizon are:

- Did animals come over the horizon and into Bronze Age societies in Rogaland?
- Did they overstep—and thus initiated greater proximity?
- Or were they brought in, coaxed in, even?
- What happened when they came?
- How did this change the lives of members of human society?
- How did this change the animals—and who were the animals?

Whereas the first three questions above are framed around how animals entered human societies, and thus came to shape the three-aisled longhouse, the three final questions probe into the nature of the ontological status that these animals came to inhabit. The first questions will be addressed by a close and comparative analysis of the data that was presented in chapters 4 and 5. Regarding the latter questions, I discuss models that are appropriate for the statuses and agency of the household members—were they stakeholders or shareholders, or should we consider humans, dogs and sheep as citizens?

Rounding up the evidence

I have presented a variety of sources pertaining to the Bronze Age practice of living with sheep, and dogs. A quick summary of the evidence that relates to the Bronze Age households in Rogaland follows:

- In the Early Bronze Age, the environmental studies show that human intervention and intentional management to facilitate grazing drastically altered the landscape. The animals were agents in this process, since their grazing kept the landscape open. This is one phase in a series of changes, starting in the Early Neolithic, with another change occuring in the Late Neolithic, and finally a phase of intensified clearance, in which the landscape had become a mosaic of land for grazing and forests.
- A majority of sheep, in addition to goats, pigs and some wild animals are present in the scant faunal data. Faunal remains from cattle are missing.
- The rock carving from Valhaug shows herding, the triad of human-dog-sheep.
- Sheep coprolites were found inside a replaced posthole in house II from Austbø lok 20, and additionaly near a wall post (presumably just outside the house).

- 11 of the altogether 16 three-aisled houses from the Early Bronze Age are interpreted as having a human living section and a byre, due to the following factors: Distribution of inner structures, like more closely placed trestles in one part of the house, inner dividing walls, the position of entrances, smaller posts interpreted as traces of makeshift pens, and floor levels being lower in one end of the house. Notably, the byres do not have traces of stalls.
- From the late two-aisled house at Talgje, loom weights were found inside the house, indicating textile production.
- The woman in the Rege burial was dressed in a corded skirt made from wool. Wool textiles were also found in the Guttormsmound burial and in the Sothaug mound, both from the Early Bronze Age (see discussion later in this chapter).

The evidence allows for moving beyond the broad animal category and look specifically at species. Although not the only domestic species in the Early Bronze Age in Rogaland, sheep have a significant presence as the most important species in three respects: first, in the faunal remains, second, in wool textiles and third, in their presence by way of coprolites inside a house. Dogs also have a significant presence, both on the rock art motif and in addition cast in precious bronze. In chapter 5 I took a closer look at sheep and dogs, which deepened the understanding of what kind of an impact an increased proximity and intimacy between humans and sheep would have upon their co-habited world. Here, I go deeper into the consequences of introducing sheep into the household.

Trust, socialization and habituation embedded in the architecture

In chapter 3, I argued that I seek to explore the effect of movements, and their inherent agency, on farming societies in the Bronze Age. I noted that animals by their agency and volatile movements transgress in space, and either intrude or flee; in the words of Henry Buller (2012)—they flirt between the worlds of lost and found. However, from a state of flirting, sheep transgressed—they came from beyond the pale, in the words of John Berger, "from over the horizon" and transgressed the human living-space. Human intentionality aside—what matters is the consequences, which is the three-aisled longhouse. Its construction is most probably a spatial negotiation to facilitate that transgression.

So, wherein lies the agency of sheep in the Early Bronze Age? Obtaining wool would require a more intense and close relationship of manageability that is highly dependent upon the development of trust between humans and sheep (see Figure 5.1). I suggest that the shift in building tradition, from two-aisled to three-aisled longhouses, might have been associated with a shift in socio-economic household practices. This shift consists of an intensification of the human-sheep relationship, where sheep were drawn closer to the household. The increased proximity could, I suggest, have led to sheep being taken into the house. But sheep are normally happy to be outside, hardy and self-sufficient, so why would such a shift happen? I propose that it is related to secondary products, probably wool and possibly also milk. Of course, these animal parts could be obtained without housing animals under the human's roof. Obtaining them requires, as I stated in chapter 5, a novel set of actions that led to particular practices. And keeping animals indoors, at least for part of their lives, is, as I will argue, beneficial for those sets of actions. It also has implications for how the relationship was allowed to develop.

With the changes in architecture, changes in ways of living occurred too, although it cannot for certain be extrapolated that the changes came about because the Bronze Age farmers wanted to change their way of living. This question is a conundrum, which is bound up with the query "Why?" Why did this change in architecture take place? Is it, as others and myself have argued, because animals were brought inside the house? What could be the reason for indoor stalling of animals, other than the economic and functional reasons that are frequently given (discussed in chapter 4), but have been largely disproved by Zimmerman (1999b)? Both Årlin (1999) and Rasmussen (1999) suggest intensification of social relationships between humans and animals (assumed to be cattle, see Holst and Rasmussen 2013) to be imperative for the development of the byre. Although I believe there is some truth to this, and that the new living arrangement certainly resulted in such deepening of relations, I will venture forth with a different explanation.

In my view, intensification of relationships is not a sufficient explanation—it is too vague, and I further believe there are particular historical situations that give rise to changes in relationships and configurations of society (see for example Prescott 2012). I therefore suggest that rather than somewhat vague explanations of increased affinity and changes in social practice, the reason underlying this change are particular practices and particular relationships that grew out of sets of actions intended to obtain wool—and drawing upon knowledge of sheep and their prerequisites, these new practices became implemented in a new way of living. Let me explain: there are some clear advantages by keeping animals inside the house. One is that the animals become habituated with people and allows for the development of mutual trust between human and animal on an intimate level that would never happen with flocks kept outside. With dogs, it has frequently been demonstrated that they have been "imprinted by people" for millennia (e.g. Haraway 2008; Clutton-Brock 1987): They have developed the ability to bond closely with people, provided they are socialized and habituated with humans during their first period of life. Socializing at this stage is crucial to develop positive feelings towards, and the ability to trust in, people (Appleby 2002; Solbakken 2010). Dog trainers encourage early socialization and holds that it is key to develop the kind of healthy social security that is so vital to create a strong working partnership (e.g. Price 2011).

As outlined in chapter 5, there are studies that demonstrate that early human handling of lambs leads to increased affinity for humans in lambs. Markowitz, Dally, Gursky and Price (1998, 580–581) found that the three first days of life are of paramount importance regarding socializing and habituating lambs to other species, and can establish so-called secondary attachments to human handlers. Ryder is on to this when he states, in relation to the domestication process, that "the key biological process appears to have been the "imprinting" on a young animal of a human being in place of its mother" (Ryder 1983, 3). However, I would argue that "imprinting" was not advantageous for the farmers in the Early Bronze Age. Imprinting would lead to a *blended flock*, in which species specificities would be lost. In a blended flock, the natural behaviour of the animals would be lost due to lambs being imprinted by humans from birth. For example, this would cause problems with the fertility of the animals, as animals imprinted by humans want to mate with humans rather than their own species when they reach adulthood (Denenberg 1962, 112).

A different way of relating to sheep must have taken place, and I suggest that rather than imprinting the lambs, humans and lambs must have been co-socialized, which would create a *shared flock*. Humans and sheep shared the house, but each had their part. A shared household would provide farmers and sheep with a shared flock, rather than a blended flock. In a shared flock both parts could keep their distinctive traits, whereas in a blended flock, some species traits might well be lost. But, in a shared flock, the two kinds would be socialized with each other, and at the same time retain their species specificities that made them both thrive and enable a healthy and fertile flock.

I suggest that one of the reasons underlying the change to three-aisled houses was to create a setting in which co-socialization, and inherently trust, could be nurtured. The way I see this tangibly happening is that at first, ewes about to give birth were taken into the house and the mother, in a social setting where humans were omnipresent, reared the lambs. In this way, being around humans would be as natural to the lambs as breathing. The lambs could become habituated to human hands as "kind hands." This would be advantageous for both humans and sheep, as the sheep could be assisted in cases of difficult births and their lambs would be more sheltered and protected than they would have been outside. For the humans this had the obvious advantage that the sheep became much more manageable and it would be much easier to pluck the valuable fleece off. This setting, then, formed the backbone of the social contract between humans and sheep.

Sheep as a species differ from dogs, that are able to retain very deep relationships to both human companions and other dogs, but without losing their "dogness" (Bradshaw 2011). Thus, dogs can form blended flocks with humans, whereas sheep cannot do that without losing their "sheepness." The ontological distinction that is observed between the different kinds—sheep, human and dog—are therefore probably based upon inherent characteristics that define their mode of agency. I acknowledge that, nonetheless, the three kinds of agents would have impacted one another, and their co-authored life-world would leave marks on each. Sheep would have, in a social, behavioural and ultimately genetic sense, become a little bit human, and a little bit dog, and so on, in order to accommodate interspecies communication.

Many that keep feral sheep (*Utgangersau*) today leave them to give birth by themselves in the pasture. Mostly feral sheep are able to give birth successfully outside, to healthy lambs. However, sometimes it goes wrong, and when lambing goes wrong and is unsupervised, it goes horribly wrong, resulting in the death of the ewe and her lambs. Some think that leaving the ewes to give birth by themselves borders on recklessness and on unethical treatment of the sheep.[1] I suspect that Bronze Age farmers did not want to risk losing any of their flock if they could help it, and were keen to supervise and aid the ewes to ensure successful births to the best of their ability. Soay sheep breeders also recommend keeping the ewes close during lambing so that the lambs can be picked up and checked immediately after birth, as soon as their mother has cleaned them.[2]

Another disadvantage to letting the ewes give birth in the pasture is that the ewe will jealously guard her lambs and let nobody come near her, or them. Keeping ewes

1. Personal comment by sheep farmer Kristin Sinnes, 15.09.2013
2. http://saltmarshranch.com/husbandry/working-newborn-lambs.shtml

Figure 6.1 A ewe stalled indoors during the birth of her lambs, enabling the sheep farmer to establish a relationship based on trust from the very beginning, in which the lambs accept her as their secondary caregiver. Photograph by Kristin Armstrong Oma.

inside to let them give birth with humans around would, on the other hand, ensure that lambs were socialized with humans, and even bond with humans (Figure 6.1). In such a scenario, when humans are birth helpers, the lambs can bond with their human "parent," identifying them by smell and sound. In cases of intense socialization, lambs can become very friendly and loyal to their human "parents," such as for example bottle-fed lambs.

The research on early socialization in sheep and dogs demonstrates that indoor stalling of farm animals, especially of the young, deeply impacts the animal's ability to develop positive feelings and trust towards other species—not just towards humans, but also towards farm dogs. As stated earlier, good stockmanship facilitates positive relationships between farmer and the farm animals in her/his charge, and makes sense regarding animal welfare (Dawkins 1980) as well as giving higher production yields (e.g. Seabrook 1972, 2001).

Markowitz *et al.* (1998, 573) claim that in modern "factory farming"—extensive range systems—the animals experience little, neutral or aversive contact with humans, "causing them to be less tame than animals cared for in more intensive systems" (1998, 573). Archaeologists should therefore avoid using "factory farming" as a model for past farming (Armstrong Oma 2013c). I claim that organic farming is better suited as a metaphor for Bronze Age farming than the extensive systems of modern "factory farming." In organic farming, some are voicing a need to look at animals as individuals (Alrøe,

Vaarst and Kristensen 2001; Lund 2002; Lund *et al.* 2004a, 2005; Young, 2003). This kind of farming creates a greater proximity with animals, and the relationship is more resilient, allowing sensory and tactile experiences to generate a more complex and conscious relationship. Farmer Rosamund Young (2003, 6–7) feels that the best way of making the most out of a flock is to treat them as individuals. An ethnographic study shows that close-up, long-term interaction with animals on a farm is likely to generate a strong awareness of individual differences between animals, in terms of appearance, behaviour, temperament—to the extent that the farmer who interact regularly with the animals will come to see each of them as unique (Knight 2005, 6; Theodossopoulos 2005). A larger work-input is required in this kind of farming (Seabrook 2001; Lund *et al.* 2004a), and necessitates greater creativity, but gives in return more stimulating challenges (Young 2003). Outcomes of this approach would be that animals are expected to live in a natural environment, and as far as possible, they are allowed to remain in it, both in life and in death, even being slaughtered in familiar surroundings.

The sheep people

The kind of human–animal relationships that are outlined concerning the Early Bronze Age in Rogaland create particular lived practices. Living with sheep, dressing in textiles from sheep fleece and eating mutton are intense multi-level embodiments of sheep. Such deep phenomenological entangling with a different species acts on the senses in many ways. One striking example is the olfactory component; smell. Smell is experienced as deeply meaningful, and there is in itself no such thing as "good" or "bad" smell, the judgement of the pleasantness of smell is culturally determined (Porteous 1985). However, going into a sheep shed full of sheep is overpowering for the olfactory senses. Sharing household and roof with sheep would lead to the particular smell of sheep becoming as familiar as one's own smells. The smells of humans, dogs and sheep would intermingle—although the smell of sheep would probably dominate, given their greater numbers. For those living in such a household, this smell would simply become the smell of home. But for people belonging to other communities, those that lived with sheep must have smelled like, and subsequently become, the sheep people.

Similar to past experiences such as the taste of food, the perceptions of vibrancy of colours and the noises of a household, the smellscape of the past can only very rarely be experienced by archaeologists. But from the presence of sheep materiality, such as textiles made from wool, inferences can be made that let us glean past sensory experiences.

Wool is a textile fibre that has exceptional qualities, and as it is "living material" that is created by individual sheep, it retains biological traces of the sheep and is formed by the sheep's life situation. Significant factors are age and gender. Throughout the lifespan of sheep, they create wool with different qualities. Lamb's wool is soft and fine but throughout the life of a sheep the texture of the wool changes. A remarkable story[3] from the traditional society of sheep herders on Shetland tells of a blind man that goes into the sheep barn (in total darkness, which of course did not make any difference to

3. Ethnographer and artist Amy Lightfoot, who experienced this, often recounts the story and Lise Bender-Jørgensen and Ellen Tjørhom Bøe told it to me.

a blind man), and just by touching the sheep and their fleece he could say: "ah, here we have a fairly young sheep, who has had lambs twice, and here is an old whether, well on nine years old" and so on. It is an established fact that the fleece of sheep changes both through the lifespan of the sheep and through the year, particularly regarding strength and fineness of fibre.[4] As such, sheep give wool that at different times in their life is useful for making different kinds of textiles. Consider, for example, the quality needed for a baby's blanket versus for a sail for a large seafaring ship. Rast-Eicher and Jörgensen (2013, 1224) explains this as follows: "They knew from experience which fleece or wool was suitable for which product. This is a form of embodied knowledge, transmitted from person to person in a form of apprenticeship." According to Rast-Eicher and Jörgensen, sorting wool was a very important job that required specialist knowledge. In addition, I add that an intimate knowledge of sheep as individuals was advantageous within the framework of embodied knowledge, and for the transformation of their fleece from a living materiality to textile. The consideration of sheep as agents—as producers—and thus a vital fundament to the embodied knowledge of human wool-workers, is lacking in studies of prehistoric textiles.

Textiles, then, are emblems of knowledge, embodiment and intimacy between humans, sheep and sheep materiality that can be studied archaeologically in wool textiles—dressing up in sheep. The textiles would retain not only the smell of sheep, but also the feel of touching the fleece. In the following section, I will explore the archaeological evidence for the Bronze Age penchant for dressing in sheep, and in the process get to know some members of the sheep people more closely.

Bronze Age individuals: Priestesses in a corded skirt and swordbearers

The cause underlying why I believe sheep became household members lies buried in a soft, warm, strong, lustrous fabric: the wool textile. The sub-text is that in the Early to Middle Bronze Age, textiles start to appear in graves, and tools consistent with textile production are found, such as the loom weights from the house at Talgje.

The loom was probably invented in the Near East in the Early Neolithic, and was used to make fabric from flax (Bergerbrant 2007, 51). The innovation of weaving was not in use in northern Europe until the Early Bronze Age, when textiles made of wool were produced. Wool textiles start to appear in graves in southern Scandinavia around the Early Bronze Age period II, attesting to an increase in wool production and technological advancements in textile production. Well-preserved wool costumes for women and men are found in Danish oak coffin burials. The most famous example is the costume of the Egtved girl (see Figure 6.2).

The Egtved girl is the remains of a young woman, dressed in a short tunic and a cord skirt, both made from sheep wool. A belt plate was fastened onto her belly with a belt. Her costume is similar to Danish Late Bronze Age figurines, most notably from Faardal and Grevensvænge (Glob 1969). The provenance of the wool fibres in her costume and blanket were described in chapter 5, and more famously, the young woman was also

4. Documented in the movie "Ullarbeid på Shetland—en video om tradisjonelt jordbruk på Shetland, basert på sauedrift" (Wool work on Shetland—a video on traditional farming on Shetland based upon sheep farming) from 1996. Amy Lightfoot was responsible for the movie, which was produced by NORTURA Environment.

Figure 6.2 The costume of the Egtved girl. http://guderoggrave.dk

herself analyzed to cast light on her life history (Frei *et al.* 2015). The results are surprising, since her body still retained traces of the life she lived, and showed that both her and her material culture experienced considerable mobility. Similar analyses done on the Skrydstrup woman (Frei *et al.* 2017b) indicates that at least some Bronze Age women have been mobile.

The woman died and was buried in an oak coffin some 3370 years ago (based on dendrochronology of the oak coffin). She was dressed in a corded skirt and a short blouse, and placed on a cowhide. At the edge of the coffin, a single yarrow flower was laid. Due to special circumstances, a layer of soil rich in iron prevented the organic material in the grave from decomposition. Not only are her costume and grave goods exceptionally well preserved, but also her skin, hair and nails. Her hair was shoulder length and blonde, and maybe partly the reason why she has become the icon of the Danish woman through history.

A team of specialists used biomolecular, biochemical and geochemical analyses to reconstruct her mobility and diet (Frei *et al.* 2015). Strontium isotope analyses of her hair and teeth demonstrates that she was not born and bred on Danish land, rather, she had travelled from afar, and also returned to her native land (or somewhere else) twice within the last year of her life. She was buried together with the burnt remains of a child of 7–8 years, who was similarly not local to the south Jutland area. Possibly the two came from the same place.

Based on geological baseline values, several places of origin have been excluded: all of Denmark, northern and middle Germany, eastern England and southern Sweden (see map in Frei *et al.* 2015, 2, fig. 1). The team suggests The Black Forest region in south-western Germany as a possible place of origin, based on the typological similarity of the archaeological artefact record patterns. However, since baseline values for strontium isotopes have not been established for Norway[5]—could it be that she originated from somewhere in Norway, for example Rogaland (Armstrong Oma 2015)? Cultural similarities between Rogaland and Jutland have been known for a long time (e.g. Hornstrup 2011; Kvalø 2002; Nordenborg Myhre 2004). For example, the wetland area of Revheim in Rogaland is the only location in Norway where a pair of lurs (blowing horns) have been found, almost identical to pairs found in bogs in Denmark. There is also another pointer towards a connection between the Egtved woman and Rogaland.

The woman buried in the Rege mound in Sola municipality, Rogaland, exhibits a great likeness to the grave of the Egtved woman, and was buried with remarkably similar grave goods (see Figure 6.3). The mound is situated where the flat heartlands of Jæren meet the sea, and was the final resting place of a woman, presumably an adult (Bergerbrant 2014, 83, with reference to a personal comment by Joakim Goldhahn). Unfortunately, only a few bones are left of the woman and she cannot reveal events from her life. The mound is in itself a massive monument, at 20 m across and a height of 3 m. In the Bronze Age it must have looked very impressive. It is dated to 1500–1300 BCE, during the Early Bronze Age II period. Unfortunately, organic material was not well preserved, but several bronze objects were found in their original position on the woman's body (Myhre 1979).

She wore some fine bronze ornaments, a collar, belt plate (tutulus), two bracelets and a spiral-ended fibula, and a dagger was placed on her hip (see Figure 6.4). In addition some costume ornaments were found, among these were small tubes that must have been fastened to a corded skirt, similar to those found in several of the Danish oak coffin graves such as the famous Egtved woman (Figure 6.3; Kristiansen and Larsson 2005, 351; Stig Sørensen 1997), as well as depicted on the Late Bronze Age figurines from Faarup (Glob 1969, 191; Kjær 1927) and Grevensvænge (Glob 1969, 192–196).

Another Early Bronze Age mound in Bore, Klepp municipality, is known as Molkhaug (see Figure 6.5). A woman was interred in the mound, and although she was not as richly equipped as the Rege woman, she was buried with a small tutulus and two fragments of bronze tubes that probably were fastened to a corded skirt. Human and animal bones were found in the grave, as well as pottery and flint (Engedal 2010).[6] The human skeletal

5. Personal communication Anne Lene Melheim.

6. Described in detail in the catalogue of the Museum of Archaeology, University of Stavanger: http://www.unimus.no/arkeologi/forskning/index_katalog.php?museum=am&id=20487&museumsnr=S6020

Figure 6.3 Illustration created by the artist Hege Vatnaland to highlight the similarities between the Rege woman (left) and the Egtved woman (right) for the Bronze Age exhibition "Where do we come from?" at the Museum of Archaeology in Stavanger, curated by the author. The Rege mound as it appears today is seen in the background, photographed by Terje Tveit / Museum of Archaeology, University of Stavanger.

remains indicate that she was an adult (Bergerbrant 2014, 83, with reference to a personal comment by Joakim Goldhahn).

The fact that corded skirts appear in different contexts suggests they were part of a set costume possibly worn by particular members of society to signify their gendered role (Bergerbrant 2007; Stig Sørensen 1997). They are sometimes referred to as part of a ritual costume (see discussion and references in Bergerbrant 2014). However, Bergerbrant's study of the corded skirts in relation to the age, social status and wealth

Figure 6.4 The contents of the female grave from the Rege mound. Photograph by Terje Tveit, Museum of Archaeology, University of Stavanger.

Figure 6.5 Tutulus from Molkhaug, Bore. Photograph by Terje Tveit, Museum of Archaeology, University of Stavanger.

in bronze of their bearers does not fully support an understanding of them as ritual garments, since there is a great diversity in the contexts in which they appear, and that they were worn by a range of social categories among women (Bergerbrant 2014). Bergerbrant suggests that the evidence points towards "women wearing corded skirts belong to a category of people that had the right to be buried in mounds, since most of the graves are found in mounds" (2014, 92). This concurs with the women buried in the Rege mound and the Molkhaug mound, which differ significantly regarding grave goods but were both granted burials in massive mounds. Construction of the mounds would have been a considerable effort, bearing in mind factors like supplying materials, the work-force in itself, the feeding of the workers, and the loss of manual labour to other tasks vital to society as long as the construction lasted.

Another dimension is added to the Rege burial by slabs inside the chamber that were decorated with rock carvings of spirals and sun symbols, ubiquitous symbols in the Bronze Age world (e.g. Kristiansen and Larsson 2005). Interred in the chamber we glimpse a woman with a special standing in society—possibly a religious specialist (Syvertsen 2003)—I prefer to call her a priestess. Remarkably, a priestess dressed in a skirt made from wool textiles.

There are two more finds of wool textiles from Rogaland, one from Sola and one from Karmøy, north of Jæren. Inside the large mound known as Guttormshaug at Reheia, also known as Blodheia, were two textile fragments, besides various bronze objects, such as a sword, fragments of a dagger, bronze buttons and a brooch (Myhre 1995, 28). Thus, like with the Rege priestess, remains of costume ornaments are found—and in this case a remnant of the wool textile was found. Another find of wool textiles comes from a large mound known as Sothaug, at Jåsund in Sola municipality. A man was buried in this mound, with princely attire, a sword and bronze buttons[7] (see Figure 6.6). Amateurs excavated the mound in 1842, and discovered a stone-built chamber, which contained an unburnt skeleton that quickly disintegrated. The bronze sword was placed on the chest of the skeleton and is dated to Early Bronze Age III period (Myhre 1979, 68, 74). These finds testify to the use of wool textiles in Rogaland in the Early Bronze Age, at least by prominent members of society.

The fact that individuals with a special position in society were buried in large mounds, with, at the time, outstanding bronze objects, and were also clothed in textiles made from wool, attest to the value of sheep wool. The woman buried in the Rege mound in her seemingly symbolically significant garment made from sheep wool is remarkable. Dressing in sheep wool is in one sense an act of embodiment, in which sheep fleece become a second skin. Sheep enveloped the wearer of wool textiles. A tactile experience, feeling, looking and smelling like sheep. Animal products put onto the body become integrated with the human body, since the body consists both of an inside, a container, and also an outside surface. Clothes and ornaments worn on the body are incorporated into a person. Clothes may have magical properties towards "making the person" (Game 2001, 4), and come to represent the person that inhabits them. This is reminiscent of Viveiros de Castro's notion of the manifest bodily form as

7. Described in detail in the catalogue of the Museum of Archaeology, University of Stavanger: http://www.unimus.no/arkeologi/forskning/index_katalog.php?museum=KHM&museumsnr=C1045&id=84516

Figure 6.6 Sword from Sothaug, Jåsund. Photograph by Terje Tveit, Museum of Archaeology, University of Stavanger.

an envelope or a clothing that conceals an inner form, somewhat shared between species (2004, 465). Could, thus, humans and sheep at some ontological level be the same?

We can envisage intimate encounters: these women were probably living with sheep, maybe working with the fleece, and also dressed in sheep. An enhancement in the interest in, and value of, the sheep fleece would lead to a more complex and intimate relationship with the "owner" of the fleece—the sheep. The sheep themselves probably responded positively to the—for them—familiar look, smell and feel of people clad in wool. From this perspective, sheep are not objects/products, they are subjects/producers. New ways of creating human-sheep relationships would be necessary to establish sufficient trust and intimacy for humans to be able to take the fleece off the sheep. Being with large flocks of sheep and managing them would require the kinds of new solutions present in the three-aisled longhouse—that changed the roles of all involved compared to previous ways of living.

Zooming out: Framing the sheep people within the wider Bronze Age world

In this book I have examined households and agents through a micro-scale lense. This approach has allowed me to access and ascribe a layer of identity not previously imagined. However, identity is multifaceted and dynamic in its nature and is made up by both the very small and the very large parts of life. In this latter part I expand the archaeological context to consider the wider Bronze Age world and its impact upon the identity of the sheep people, especially focusing on its power structures, long-distance networks and ritual communication. In the following, I explore some general features of the furnished Bronze Age landscape such as settlement patterns, grave mounds and rock art, as well as zooming out and framing Rogaland in the wider Bronze Age world.

In addition to the farms where people lived, they were also embedded in a wider landscape of varying topography, geology, soil conditions, elevation above sea level, and so on. From the Late Neolithic onwards, the landscape was dotted with farms made up of single two-aisled houses (Soltvedt *et al.* 2007). Coils of smoke would have slowly unfurled from the cooking fires and the farms would thus be noticed beyond their visual presence, but also by their sounds and their smells. Already at this stage, the landscape was marked by the agrarian contract between people, the land and the animals.

With the onset of the Bronze Age, new kinds of landscape furnishings came into being, both in the form of large grave mounds and in the form of rock art, both are most often interpreted as ritual landscape markers (Austvoll 2014; Nordenborg Myhre 2004; Sørskog 2007; Syvertsen 2003, 2005). The region of Jæren, and also further north along the coast, the large island of Karmøy, still retains visible monuments that marks the graves of prominent members of the Early Bronze Age society. In Rogaland, the mounds appear during the Bronze Age I–II period, at roughly the same time as the three-aisled longhouses start appearing (Austvoll 2014; Nordenborg Myhre 1998, 2004; Syvertsen 2003, 2005). The early mounds are massive monuments, their size is often accentuated by their strategic placement in the local topography, on natural hilltops. They are also often placed with regard to visibility, not only from the surrounding areas on land but also from the sea (Nordenborg Myhre 2004; Syvertsen 2005). The mounds that have been excavated show the graves of individuals, both women and men, placed on their backs with status objects made from bronze. The placement of the dead and the ornaments point to a desire to display wealth, or alternatively, perhaps, embodied objects that reflect aspects of the identity of the deceased. Women were buried with richly ornate jewellery such as belt plates, tutuli, bracelets and neck collars, and often a dagger (Myhre 1979; Syvertsen 2005). The bronze ornaments must have dazzled the onlookers when the women moved and the sun caught them (Kristoffersen and Armstrong Oma 2008). Men were buried with weapons, swords and daggers. Some of the graves hold chambers with slabs with rock carvings (Syvertsen 2005). The Rege burial, discussed above, is one representative of this burial custom.

These novel landscape features were a part of the life world of the sheep people—and beyond a passive backdrop, these monuments were actively created and maintained by the sheep people. For example, the island of Åmøy in the Boknafjord just north of the Jæren region, and near Hundvåg, is home to one of the largest clusters of rock art in Northern Europe. The figures range from sun symbols, some large fish believed to be life-size halibut, foot soles and ships, the largest one being 5.5 m long (Fett and Fett 1941). Åmøy is sometimes referred to as the island beloved by the gods. The foot soles are interpreted as marking the presence of divine beings, and together with the absence of settlement traces, this could indicate that the island was off-bounds for human occupation, and should only be visited on special occasions (Kristoffersen and Armstrong Oma 2008). No settlements dated to the Bronze Age have so far been discovered on the island.

This site reflects the religious beliefs in the Bronze Age, in which the cosmology and celestial beings, like the sun, were structuring principles that ordered the world (e.g. Kaul 1998, 2004). In the religious universe of the Bronze Age, the sun was central to the cyclical understanding of the world. The journey of the sun through night and day and through the year was possibly tied to Indo-European ideas (Melheim 2004, 2006) that reflected a belief in the reincarnation of the soul (Armstrong Oma 2008b; Armstrong Oma and Melheim In prep; Kaliff 1998, 2005). Archaeologist Flemming Kaul (1998, 2004) has systematized motifs on rock art and objects, particularly razors. He postulates that in the Bronze Age, there was a belief in the sun as the centre of the universe, and that all benevolent forces helped the sun on its journey through day and night. Through the night, the sun sails on the night ship, known through rock art motifs of ships holding sun symbols. In the morning, the sun rises from the sea, possibly helped by the powerful halibut? According to the myths of the journey of the sun, fish bring the sun up from the sea at dawn, from the night ship into the morning ship. As mentioned above, there are large carvings of fish, probably halibut, on Åmøy. The halibut is a large, strong fish, and if any creature in the sea would be sufficiently powerful to help the sun rise from the sea it would be precisely this one, one of our most myth-laden creatures. In earlier times, it was seen as a holy fish, *hali* is derived from holy.[8] Halibut can grow to 3 m in length, and weigh up to 300 kg. It lives in deep waters, but will sometimes come to the surface during the summer. Halibut are round like the sun, and like the Trundholm chariot sun disc, it has a dark and a light side. Fishermen say that when hauling it in, a mystical light would come from the depths, probably because of its large, light underside. The behaviour of the fish was presumably familiar to Bronze Age people living by the sea in Rogaland (Kristoffersen and Armstrong Oma 2008).

In the morning, as the sun rises from the sea, and the night ship is pulled to the surface by the halibut, the morning horses take over, they gallop across the heavens and through the day. As evening falls, the horses land by the ship, and helped by snakes and fish, the sun enters the night ship and travels through the sea. Rock carvings with ship motifs and cup marks and circles, interpreted as sun symbols, are very common in Rogaland (Kristoffersen and Armstrong Oma 2008; Mandt and Lødøen 2005, 231–241). At the spectacular rock art site of Fluberget, by Revheim in the northern part of Jæren, ships seem to sail into a crevice in the rock. A constant flow of water trickles across the crevice so that the ships sail into wet darkness (Nordenborg Myhre 2004, 108, 178). Horses are not frequent but they are found. At a panel at Ølberg in Sola municipality, horses, ships and a large circle appear together with cupmarks (Hernes 1992).

8. In Norwegian halibut is called *hellefisk*, "helle" is derived from "hellig," which means holy.

The imagery of Bronze Age rock art brings us close to the carvers in the deep past, and goes some way towards bridging the time gap between now and then, since they communicate directly by visual and tactile experience. But still, the motifs lack narrative structures and are not always easily deciphered—they remain enigmatic. The relationship between humans and animals has been framed within the wider understanding of Bronze Age cosmology, expressed as, among other aspects, this understanding of the journey of the sun, but also by animal figurines, ritual depositions and grave goods. Meetings between humans and animals were surely important to Bronze Age cosmology as expressed on the medium of rock art (e.g. Mandt and Lødøen 2005), both in expressions of hybridity and in the form of encounters. Subsequently, the animals on the rock art panels surely also represent real, living animals. The rock art from Valhaug (see Figure 1.1) is unique in that it lets us get close to a situation that must surely reflect a lived reality, with real, living agents, and the flow of energy that make up their dynamics.

Animals in grand and small narratives

The short overview above, based upon rock art, reflect the role of animals in the "grand narrative" of the Bronze Age. This way of perceiving animals is in stark opposition to the discourse surrounding settlements and byres, in which animals are considered in purely functionalist terms (although there are exceptions, see Årlin 1999; Rasmussen 1999). I have previously suggested that a common ground could be identified where the animals of the mythological sphere and as participants in economic strategies meet, namely the social practice of the household on the farm (Armstrong Oma 2012).

A model originally developed by Larsson (2002)—inspired by Mary Helms (1988, 1993, 1998)—illustrates the "grand narrative" and its structures underlying the Bronze Age world-view and religion, according to Kristiansen and Larsson (2005). This model is particularly salient since it resonates with the animals in the postulated myths, in which animals carry the sun and are incarnations of gods. Animals are placed in the sphere of otherness and ancestors—out there beyond, the sphere that is the aim of the journey. An essential problem regarding the model is that it does not fully account for how the animals come to be in the sphere of otherness, which animals have this kind of role and what the animals are actually doing in that sphere. Larsson mentions iconography where horses pull the sun and also the presence of various animals on rock carvings—but he doesn't discuss this any further (Larsson 2002, 101).

The close physical relationship between humans and animals is demonstrated to be a recurring theme in Bronze Age depictions and shows how animals are useful in religious beliefs both as the incarnations of the gods, as their messengers or other assistants, and as agents with decisive powers in the mythologies. Animals within iconography considered to depict mythology are placed in the realm of the heavens or the sea, far from earth and daily life. In Bronze Age discourse, they are seen to belong to the cosmos, as central agents in the grand scheme of things. Kristiansen and Larsson (2005, 320) suggest "[i]n the pre-modern perception of cosmos animals and nature are powerful and equal to Man. Animals and nature took part of the creation of the universe, and the difference between them is constantly fluid. Animals become humans and vice versa." However, what "equal" signifies in this context remains somewhat enigmatic, and unexplored.

It may go beyond the realm of cosmology. This exemplifies that within the "grand narrative," the cultural backdrop to the connection between humans and animals is not accounted for, other than stating that humans are in need of animal powers, a rather simplistic and almost totemic statement. In what way animals are important, other than as a source of desirable power, or even what this power consists of, is not explored.

The roles of real-life animals in real-life households, beyond the realm of myths and gods, remain obscured by the emphasis on cosmological institutions. In my view, one of the main shortcomings of the "grand narrative" is that it fails to create a connection between everyday life and cosmological institutions. Yet, this perspective has important implications regarding animals. Transposing animals into the sphere of otherness has consequences for their ontological status, particularly for how they are integrated within the "inside," in the household/settlement context. As a whole, the model forms an important antithesis to the household, since the same animals participated in both spheres.

Wool production and wool textiles cannot be seen in isolation from these larger trends in society, identified in the "grand narrative." The results from the analysis of the textiles in the Egtved burial, outlined above, demonstrate how sheep wool from geographically distant locations was mixed to create textiles. Thus, the wider network that made up the Nordic Bronze Age culture and connected it with a wider, pan-European network (e.g. Kristiansen 1998) is also a factor when it comes to the distribution of sheep wool.

A notable trait in the mythological iconography is the absence of sheep and dogs. The motif from Valhaug is an exception to the norm, and sheep hardly appear in such contexts, and dogs only very infrequently. Why is this? I suggest it is because they were not construed as the kind of mythological "Other" postulated within the grand narrative. Their role as household members, and as producers and workers within the socio-economic contract of the farm, would give them an ontological status that was different from the mythological, deified animals of the heavens and the sea. Their physical proximity to humans, the co-authored practices they engaged in, and the interchangeability of sheep materiality between humans and sheep are all factors that could account for the social nature emerging from their mutual relationship.

Trusting textiles

With the advent of textile production, sheep must have changed social status and taken on an, until this moment, unprecedented importance. The secondary products revolution raises an important question: how to hand-pluck the fleece of a feisty sheep? Or, in other words, what implications would the use—following the procurement of—secondary products have for specific practices of a joint human–animal world? Clearly, a greater involvement, greater proximity and physical intimacy resulted. These factors would necessarily lead to an intimate knowledge of the animals in joint households. In this way, implications of economically based choices would lead to specific social entanglements. In Rogaland, entanglements between sheep and humans led to shared flocks in which all kinds of flock members lived under the same roof.

Using secondary products requires a more intense and close relationship of manageability that is highly dependent upon the development of trust. A different way of relating to sheep must have taken place. I suggest that one of the reasons underlying the change to three-aisled houses was to create a setting in which this trust could be nurtured. Humans and sheep shared the house, but had each their part. As I argued above, a shared household would provide farmers and sheep with a shared flock. In a shared flock, species would be habituated and familiar to one another, they would interact but each species would mainly relate socially to their own kind.

However, dogs might have lived in a different kind of flock together with humans, more closely resembling a blended flock. In a blended flock, social interaction would be species specific to a much lesser degree; living spaces would be open and shared between all members. Modern pet keeping exemplifies blended flocks, in which pets have the full range of the house. But, in a shared flock, the different kinds would be socialized with each other, and at the same time retain their species specificities that made them both thrive. Interestingly, one of the very few animal species that are able to live in blended flocks with humans without being imprinted by humans and develop sexual dimorphism are dogs (Bradshaw 2011). This capacity to bond so closely with humans without losing their particular dogness would go some way towards explaining the "fringe" status of dogs.

As discussed in chapter 5, there are three kinds of beings within the social setting of this shared flock, since management strategies included dogs, judging by the rock art panel from Valhaug (Figure 1.1). We therefore see a triad—three kinds of being that come together as working partners—and the relationships between humans and these animals were formed in their particular meetings, by way of mutual action, decision-making, performance and becomings. Animals and humans transgressed each other's life worlds, and became entangled in a social situation. However, the three kinds were not blended, rather they retained their species specificities and had different roles to play. Still, both sheep and dogs were probably dependent upon being raised in an environment with a strong presence of humans in order to be able to partake in such a shared flock, which depended upon mutual trust.

In chapter 5, I explored the advantages of rearing young dogs indoors. Studies of dogs and their environment as puppies shows that pups that were reared indoors with humans were much less likely to show fear and aggression as adults. Dogs that were reared outside of the house in barns and sheds and so on, had a 68% greater chance of becoming problem-dogs. Early socialization is key to the development of trust between humans and animals, and there is a window of opportunity at a very early age when the young are able to accept other beings into their comfort zone.

Farmer Rosamund Young told me that she had once taken on a lamb from a friend, and the friend drove the lamb to her house when the lamb was just a few hours old. Then the lamb didn't see this man for a very long time, but one day, a few months later, the man came to visit. And to everyone's surprise, the lamb ran straight into his arms with a great show of affection, because, seemingly, the lamb recognized his voice. Young told me it was remarkable. Biologists most often dismiss these kinds of stories as anecdotes, without scientific value. Sure, this experience is probably not something that could be reproduced in a lab. However, as explored in chapter 2, the biologist Marc Bekoff has the courage to say that the plural of anecdotes is data. But we can rest assured that Bronze Age farmers didn't think like scientifically minded biologists. They must have trusted their first-hand observations and their instincts to guide them, and having lived with dogs for a long time, they probably understood the value of rearing young animals under their roof to develop a working, social relationship free of fear and aggression.

In this sense, the agency of sheep rests in their ability to respond, to either act back or to cooperate. Humans in the Bronze Age went out of their way to create conditions that made it favourable for animals to reciprocate and cooperate. They understood the value of creating kinship ties—even across species. I suggest that the three-aisled longhouse grew out of a wish to create and develop the shared flock. It was precisely because humans saw animals as agents—beings with agency—that they wanted to create a favourable environment where they could gain the trust and cooperation of the sheep and become socialized together. Although ultimately the motif may have been economic gain, perhaps that could not be met without first entering into a social contract with the sheep. The contract entailed mutual trust and an ethics of care, it required humans to care for their sheep in the way the sheep—and particularly the lambs— needed. To fulfill the contract, the humans needed to be there for the sheep from birth onwards, to ensure a good working relationship. In social terms, this contract meant sharing their living space together with the sheep. In economic terms, it meant that sheep would let the fleece be taken off them and let themselves be milked, so that their secondary products could be used by humans.

Deconstructing models: Economic strategies and their philosophical underpinnings

The model I have outlined above is not consistent with the general presentation of human–animal relationship in archaeology. Narratives of past farming represent one of a number of areas in which archaeology unwittingly contributes to legitimize present-day ideologies (Armstrong Oma 2013c). Although much less debated than for example postcolonial narratives and political uses of the past to naturalize current day racism, it is no less important.

Modern farming has had an impact upon the epistemology of archaeology, stemming from the days of functionally-minded positivist processualism (e.g. Binford 1981; Higgs 1972) (for critique see Grant 2002, 17). It has a bias towards cost-efficiency and maximization strategies with a single-minded focus on economic gains. The rationale of the model rests upon a short-term perspective, underlying most farming strategies in the Western world (Alrøe 2001; Dabbert 2004; Lund 2002; Lund *et al.* 2004a, 2004b; Lund and Röcklinsberg 2001). Intensified agriculture has a high yield due to use of pesticides and fertilizers, and is a considerable source of pollution and is ultimately a threat to biodiversity while depleting the soil of its nutrients (e.g. Dabbert 2004; Lovelock 1992 [1991], 156–159; Mazoyer and Roudart 2006, 435–437). In this respect, we need to critically evaluate and deconstruct the models we use to talk about past societies. A long overdue example is past husbandry practices that frequently are understood through presentist models based upon *a priori* assumptions. Modern farming constitutes one of the most common models, and I wish to identify and deconstruct the foundation of this narrative.

Regarding husbandry practices, people in the past are portrayed as being driven by maximization strategies similar to today's agricultural aims (e.g. Zimmermann 1999b).

In archaeology, ambitions for the outcome of farming practices are frequently projected onto past societies. For example, very little consideration for animals is seen, and no affinity with animals is recognized. Animals are meat and calories (Albarella 2001; Grant 2002; Hamilakis 2001). Zooarchaeologists, or even zoologists, normally do the study of animals in the past, and subsequently also study human–animal relationships. The sub-discipline zooarchaeology has been instrumental in developing ways of studying animals, regarding issues such as nutrition, disease, traction, and taphonomy (see the in-depth discussion in chapter 1). Zooarchaeology focuses on identifying faunal remains that give information about the economic uses of animals as described by Whittle:

> [i]nnumerable specialist animal bone reports (to which no disrespect is intended) principally establish species present, and patterns of age and sex, with some detailed attention also to taphonomical processes. (2003, 82)

These studies are valuable, as they have shown that there are several economic strategies that can be applied to achieve a viable result.

However, zooarchaeology is concerned with one factor above all others: animals as outcome, which is seen as a one-sided focus on economy that frequently fails to take social factors into account. This mind-set is inherited from Binford's (1978) position that without exception, all the documented variability in faunal assemblages is directly referable to functional variability. The zooarchaeological economic model falls under the general critique formulated by Knight (2005, 5) where he warns of collapsing outcome (meat) and process (the protracted relationship of nurturance and care prior to slaughter) into counting calories, when these in fact are two separate dimensions of the human relationship to domestic animals. In this book I have focused on the process and its implications for how the farm was organized.

Rather than seeing past and present farming practices as embodying the same understanding, we should consider changes in terms of engagement throughout history. The historian Keith Thomas (1984 [1983]) illustrates this by his study of the English countryside; that human-environment relationships have been subject to changes in terms of engagement, expectations and attitudes throughout history. The 16th century Tudor plight towards breaking new land and cultivation was challenged by the 19th century Romantic's idea of the conservation of the wild.

Still, in archaeology, changes in terms of engagement are rarely considered beyond the moment of domestication. For example, Ingold's (2000a) postulated shift from a relationship of trust between hunter and prey, to one of the farmer's domination over domestic animals (for critique of Ingold's stance, see Knight 2005; Armstrong Oma 2010), is often taken at face value (for example Denham and White 2007, 9–11; Wengrow 2006, 81). However, greater complexity in the domestication process is increasingly claimed (see Dransart 2002; Jones and Richards 2003, 50; Whittle 2003, 80). Further, changes in terms of engagement followed within various domestication regimes as the process unfolded. Additionally, different forms of engagements at the moment of domestication would follow in different historical situations and with the variety of species. For example, the domestication of the llama would have been very different to that of cattle—or its wild progenitor, *Bos Primigenius*.

Unbinding the domination discourse by deconstructing economic models

A useful point of departure when investigating alternative models is to take a closer look at social economic theory in order to understand these changes in terms of engagement. The importance of deconstructing economic models is related not only to how choices are made, but also to how they are justified. In neo-liberal economics, the model used by traditional archaeology and zooarchaeologists is known as the "shareholder value model." This model aims at enhancing profit for the owners from a short-term perspective (Almond *et al.* 2003). It can be compared to the stock market, in which individuals can buy shares in businesses but have no personal, non-monetary investment that will endanger them if the company goes up or down in value on the stock exchange. For example, the model justifies a perspective upon the environment as a limitless resource that can be exploited with no consideration for the consequences of endangering species or even entire ecosystems.

An alternative model is the stakeholder value model (e.g. Donaldson and Preston 1995; Freeman 1983). To be a stakeholder means not only having stakes in the outcome of a venture, but further to be directly influenced by the outcome of events. As such, stakeholders have their feet firmly planted in the midst of their venture and will be personally implicated in gains or losses—they are not separate from the results. Farmers are stakeholders, as they will starve if their harvest fails, and so are domestic animals on the farm—they also will starve. This model is therefore driven by the return of society as a whole, within a longer-term perspective. Society might be defined in a very broad sense, including both human and animal members, and the environment. The environment in terms of ecosystems will suffer if the farming practice for example introduces harmful chemicals to the water or the soil.

The stakeholder economy is driven simultaneously by collaboration and by competition, which makes it comparable to smaller-scale household economy. Rather than driven by a short-term outcome for individual owners, it is driven by trust and commitment and a gain for all members of society (e.g. Hosmer 1995; Jones 1995). As such, it is analogous to what is known as an eco-contract in organic farming, a contract between humans, domestic animals and the local environment (Alrøe 2001; Dabbert 2004; Lund 2005; Lund *et al.* 2004a; Lund and Röcklinsberg 2001). Further, it is comparable to a social contract (see Armstrong Oma 2007, 69–71; 2010), and could grow out of love for the environment and nature in which people live and thus perceive directly (Milton 2002). From this point of view, the shareholder perspective advocated by zooarchaeology lacks an understanding of complexities that can govern human–animal relationships in a stakeholder model.

The tradition of inferring husbandry strategies and aspects of the human–animal relationship in settlement archaeology has traditionally focussed on what is termed economy (e.g. Earle 2002; Tesch 1992) but is really a shareholder model. Thus, the Bronze Age economy discourse is deeply implicated in the critique presented above. Moving domestic animals (termed "livestock" in this kind of discourse) into the house has frequently been explained in terms of an increased control over the animals as well as the assumed functionalistic benefits outlined above. The underlying domination discourse has its roots in the traditional way archaeologists have configured the concept of domestication from a biological and economical perspective. One salient example is Juliet Clutton-Brock's definition from her edited volume *The Walking Larder,* where she puts it thus: "a cultural and biological process [...] that can only take place when tamed animals are incorporated into the social structure of the human group and become objects of ownership" (Clutton-Brock 1989, 7). She goes on to say that domestic animals are "bred in captivity for purposes of economic profit to a human community that maintains complete mastery over its breeding organization of territory and food supply" (see also Russell 2012). Several archaeologists find support for this way of perceiving human–livestock relations in Tim Ingold's influential narrative "From trust to domination" (Ingold 2000a; for critique see Armstrong Oma 2010; Knight 2005). This discourse is bound up with a progressive view of human history as unidirectional, in which both the environment and animals were perceived to enter increasingly exploitative relations with humans (see also Cassidy 2007, 5–6).

However, as Rebecca Cassidy perceptively points out, other voices have chosen a different approach regarding understanding domestication, most notably so Steven Budiansky, who has swung the pendulum and credited animals and plants with the domestication of humans (Budiansky 1992, 1994). Other researchers have chosen a middle ground, and emphasize mutuality, fallibility and chance in the development of humans and their environment (see for example O'Connor 1997; Rindos 1984; Zeuner 1963). Natasha Fijn points out that domestication of animals in terms of "making the animals into a being of human design, is likely to be a Western concept" (Fijn 2011, 18). In her work with Mongolian herders, she has come to a different conceptualization of human–animal relations, and refers to this relationship as "co-domestic," defined thus: "the social adaptation of animals in association with human beings by means of *mutual* cross-species interaction and social engagement" (Fijn 2011, 19, original emphasis). She thereby avoids the connotations that act as a subtext in for example Ingold's (2002a) outline of domestication, namely restriction within a captive environment of a "walking larder"—i.e. animals as economic commodities. In terms of Bronze Age Rogaland, Fijn's concept of "co-domestic" is meaningful. The lack of fencing and divisions of the landscape point towards an untrammelled herding environment, reminiscent of Fijn's experience in Mongolia, that she describes as "herders and herd animals live with each other in a shared landscape, inhabiting a co-domestic, ecosocial sphere" (2011, 19). This is also in accord with Ryder's account, in which care and protection matter, rather than control and restraint. Thus, domestication does not necessarily equal domination, and likewise, to act is not to master (Law and Mol 2010)

Kymlicka and Donaldson introduce an interesting model in their book *Zoopolis* (2011). They suggest that animals should be given an ontological status as citizens, meaning that they have the same rights and responsibilities within a society that humans have. This idea relates to contemporary society, and is a contribution to the animal rights debate. Obviously, Bronze Age society was nothing like the modern Western world, and in many ways the concept of citizenship is nonsensical from a Bronze Age perspective. For example, nation states, borders, passports, etc did not exist in prehistory. Still, the basic tenets of this idea are worth considering regarding the ontological status of domestic animals in the Bronze Age.

There is a sheep in my house

Let me sidetrack here for a moment to contextualize the current gaze at animals with that of the past. During the seminar "House of Beast," hosted by the art foundation Meadow Arts, artist Robert Davies talked about his huge canvasses of cows drawn in real-life size. The cow Buttercup is one he has captured beautifully. She is a cow from a sanctuary, and he tried to draw her as truly and utterly as he could, and he explained that his purpose for doing so was that people would truly *see* her, so that he could translate some of her essence onto the canvas. As he says, he is "working to the scale and grandeur of the traditional British equestrian portraits; (my) decision to represent cattle in this way questions our unequal responses to different types of animals, such as those used in food production."[9] It struck me while listening to his lucid talk that his desire to make people see this cow was something he need not have worried about in past farming societies. People and sheep in Austbø house II would certainly *see* each other. The question of seeing or not seeing is in this context in essence a question of allowing the other to become a subject. Members of a household recognize each other as subjects, but in our modern culture the average consumer would not recognize the subject status inherent in a piece of meat bought in the supermarket. Robert Davies invites us to to reconsider the different ways that humans regard the animals that they keep and utilize.

His intention is to reinterpret our relationship to these animals through the equal representation of the animals—racehorses as well as cows, disregarding the status differences of their usual roles. His mission is aligned with the purpose of the interdisciplinary field of Human–Animal Studies. My quest is similar, in that I invite a novel direction in the archaeological discourse surrounding animals since I try to create methods and theoretical frameworks that allow us to see the animals of the past, to subjectivize them in the same way we should seek to subjectivize the humans whose lives we study. Since people in the past clearly had relationships with animals in a multitude of ways, we are missing out on a large part of their life-worlds by not seeing the animals. Argent (2011) refers to Schwabe's (1984, 53) argument that the first sheep herders were "humanized" through their involvement with sheep; that keeping them fostered feelings of "gentleness, caring, compassion, responsibility, [and] nonviolence." The implications of this is that we need a whole different kind of discourse surrounding studies of domestic animals in past societies. To elucidate this point I am tempted to quote author Antoine Saint-Exupéry's fox in *The Little Prince*:

> "To you I am nothing more than a fox like a hundred thousand other foxes. But if you tame me, then we shall need each other. To me, you will be unique in all the world. To you, I shall be unique in all the world... If you tame me, it will be as if the sun came to shine on my life. I shall know the sound of a step that will be different from all the others. Other steps send me hurrying back underneath the ground. Yours will call me, like music, out of my burrow." (Saint-Exupéry 1995 [1943], 77–78)

This fox has nailed it, entering into such relationships means an ethics of care, a mutual responsibility, the obligation to respond and reciprocate to one another because the subjectivity and agency of the other is recognized—and they come to matter to one another. But this also means that whereas some animals choose / are cho-

9. http://robertwdavies.uk/Animals

sen to enter such relationships, others are not chosen, and these animals become the "Others," and are considered dangerous, unwanted, predators, food, and can be killed for sport.

The sheep people of today

Do the sheep people of the Bronze Age have their contemporary equivalent? Somewhat, yes. Michael Ryder's account demonstrates a deep, historical continuity in traditions and practices concerning sheep farming (1983). Whilst doing research for this book I have come upon some of them, sheep farmers that are very devoted to their flocks and their herding dogs. Without exception, the ones I have met have not been able to live off the sheep but have had to diversify. Recently, James Rebanks (2016) hit the bestseller lists with his memoirs of being a sheep farmer in the Lake District. Similarly, David Kennard (2004, 2005) has made great success with his books and TV-series about life on his farm in Devon. Sheep farmers in Norway are similarly devoted to their flocks, and will often sleep over in the sheep shed during lambing to be at hand around the clock, the family taking turns watching over the expectant ewes.[10] One farmer I have met even built a little cubbyhole and moved into the sheep shed throughout the lambing period.

During lambing, the farmers make sure to check on the lambs immediately after birth, touching sensitive body parts such as the feet and the belly (Figure 6.7). A sheep farmer told me that this is a tacit, embodied practice, she had never given any thought to why she did this prior to our conversation about early socialization in lambs.

Similarly, transhumance is still practiced, with large flocks herded to pasture in the mountains, checked regularly over the summer, and gathered at autumn. Festivities and social gatherings normally take place at these events. The sheep people who meet up at such events are highly enthusiastic about sheep and will discuss innovative farming technologies at great length, as well as entertain each other with stories of individual sheep and their antics.

So why is it that farmers choose to be sheep farmers when it is no longer a sustainable lifestyle? To be sure, tradition, social gatherings, the topography of their land and environmental conditions are important factors. But is there something beyond this, a certain *je-ne-sais-quoi* about sheep themselves that create modern sheep people? None of the farmers I have met have answered this question to my satisfaction. "Because they are nice," "because they give a variety of secondary products," "because of the wool" are common reasons that are given. Clearly, they also find great joy in working with, training, trialling and socially engaging with the dogs. But maybe also because sheep exude calmness? Like with all animals, there is the oxytocin effect (see chapter 2), but this does not account for the specific choice of sheep over other domestic animals. There has to be something special about sheep—at the very least, they have shown remarkable adaptability as a global species, going from being a species that lived in a narrow ecological niche in the fertile crescent, some 11,000 years ago, to a breed that is found all over the world. Today, they produce wool that is traded globally and star in everything from parables in the Bible to nursery rhymes to modern bestsellers, such as James Rebanks' memoirs.

10. Personal communication Kristin Sinnes.

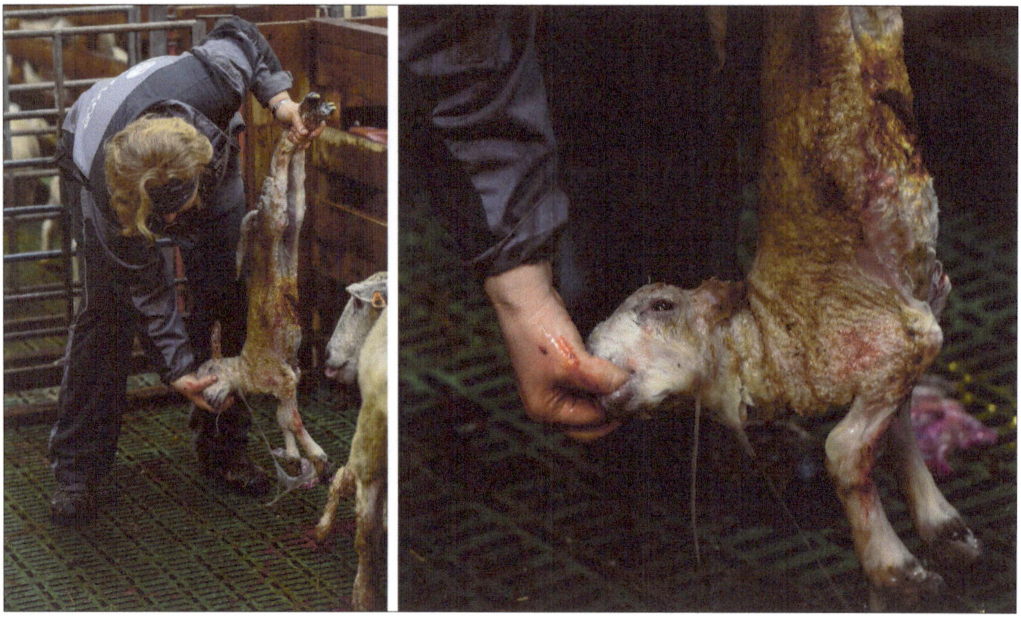

Figure 6.7 Immediately after birth, sheep farmers check that newborn lambs are healthy and can breathe. Simultaneously, the touch and smell of the farmer habituates the lambs to humans as secondary caregivers. Photograph by Anne Helen Robberstad, 2016.

Lives in the Bronze Age

At the end of this journey, can we say anything about the nature of these lives that have been under scrutiny in this book? Academic lingo aside, how did they live? Did the herders treat the sheep well? And the dogs? Francis Pryor speculates on this in his work on the Fenland Bronze Age farmers, and although we can never know, he says:

> For what it's worth, I have a gut feeling that the Bronze Age farmers of the Fens, knew their job well and did it to a very high standard. I find it hard to reconcile such a practical, efficient (in modern terms) approach to farming with wholesale, or recurrent, cruelty, malnourishment or abuse. (Pryor 2006 [1998], 101)

Pryor puts his finger on an important factor, namely that it makes economic sense to treat the animals well. Our forebears keenly observed the world around them and probably understood much the same as the cutting-edge research on animal ethology and Human–Animal Studies makes apparent. They understood that the advantage of good animal welfare was that it made sense and made life easier, the flock would thrive and all stakeholders would benefit. An interesting point of comparison is modern dog training. Today, most dog trainers will encourage positive reinforcement, not only because it makes the dog handlers feel good about themsleves rather than being a punishing sadist, but mainly because it works so much better, the dogs learn much quicker and a relationship based upon mutual trust and respect is one that is far more productive and enjoyable than one based upon fear and punishment (Bradshaw 2011).

In this book, I have explored Early Bronze Age societies in Rogaland from a perspective of the farm as an arena that is co-created and co-authored by humans, sheep and dogs. I have considered the protracted process of care and nurture prior to the slaughter. This was only possible because I took all stakeholders seriously. The final point I wish to make is that a stakeholder perspective allows recognition of the agency of non-human stakeholders—animal species are also stakeholders. In addition, animals have the ability to socially engage with humans, and to reciprocate. They also set particular ramifications for their social interaction with humans, and thus negotiate the nature of relationships, or social contracts (Armstrong Oma 2010). This demonstrates that animals can be willing—by their agency—to respond in positive ways, and to cooperate with us. They perceive us as others, as others that hold significance, as others that they trust. Obviously, this trust can be abused, and by brute strength, animals can be subdued and dominated, like women and children have been subdued and dominated in particular historical situations. Nonetheless, we should not let the possibility of domination exclude the fact that we as humans can find good ways of meeting animals, and they us. That we can create homes together, and afford each other different ontological statuses that take into consideration the particular things that each can do, and the capacities that each holds. Finally, by truly *seeing* sheep, dogs and humans, and minding—both by attributing cognitive capacities to and by being mindful of—those beings from the past, we also see their agency, and once we have seen it, we cannot un-see it. It is a factor that must be taken into account, also in telling the tales of the past.

Appendix

A summary of the 24 houses considered in the text, in which measurements and features are given. Only complete and radiocarbon dated houses have been included.

Key: WP = well preserved; P = preserved, PP = Partly preserved; RP = Replaced posts; SF = Several phases; M = Mid; OC = Off centre; PR = Pair; O = Opposing; AP = Asymmetrically opposed

House	Date	Type	Orientation	Length metres	Width metres
Østabø 43/1, Talgje, Finnøy	EBA I	Two-aisled group 1 (possibly Hybrid group 4)	W–E	13.0	6.5–6.8
Bjorhaug 25, Bø, Hå, house 3	EBA I	Two-aisled group 1	NW–SE	16.8	6.3
Nordheim 19/11, Kvåle, house 1	EBA I	Two-aisled group 1	NNW–SSE	23.0	7.2
Tjora 10/5,17,19, Sola, house 12	LN/EBA I	Two-aisled group 1	NNE–SSW	10.2	4.3–4.7
Tjora 10/5,17,19, Sola, house 9	EBA I	Two-aisled group 1	NW–SE	22.3	6
Tjora 10/5,17,19, Sola, house 11	EBA I	Two-aisled group 1	NNW–SSE	22.7	6.3
Tjora 10/5,17,19, Sola, house 10	EBA I	Hybrid group 4	NNW–SSE	15	5.5
Tjora 10/5,17,19, Sola, house 7	EBA I	Hybrid group 4	WNW–ESE	11.8	5.5
Nordheim 19/11, Kvåle, house III	EBA I	Three-aisled group 3	NNW–SSE	22.0	5.7
Austbø 7/7, Hundvåg, house 2	EBA II	Three-aisled group 2	N–S	23.0	7.75
Myklebust 3, Sola, house VIII	EBA II	Three-aisled group 3	NNW–SSE	8.7	5.6
Lunde, 4/1,13, Hundvåg, house 1	EBA II	Three-aisled group 2	NW–SE	12.5	7
Myklebust 3, Sola, house XV	EBA II–III	Three-aisled group 3	E–W	17.3	6.1
Tjora 10/5,17,19, Sola, house 4	EBA I–III	Three-aisled group 3	NNW–SSE	20.6	6.2
Berge 37/5, Forsandmoen XLV/45	EBA II	Three-aisled group 3	WNW–ESE	13.0–14.0	6.45
Tengesdal 81/4, Bjerkreim house 3	EBA II	Three-aisled group 3	NW–SE	20	5.7
Skeiane 40/ 246, Sandved house I	EBA II–III	Three-aisled group 2	W–E	ca 13	ca 8
Austbø 7/7, Hundvåg, house 1	EBA II–III	Three-aisled group 2	WNW–ESE	ca 22	ca 7.0
Berge 37/5, Forsandmoen XC/90	EBA II–III	Three-aisled group 2	WNW–ESE	16.5–17.0	7.0
Berge 37/5 Forsandmoen XXXIIB/32B	EBA II–III	Three-aisled group 2	WNW–ESE	21.0–22.0	7.8–7.9
Berge 37/13-14, Forsandmoen CCXII/212	EBA III	Three-aisled group 3	W–E	15.8	6.4
Berge 37/5, Forsandmoen LIX/59	BA II–V	Three-aisled group 2	NW–SE	22.0	7.6
Kleppe 1/6, Kleppevarden house I	BA II–IV	Three-aisled group 2	NW–SE	ca 28	ca 7.0
Kleppe, 1/22, Kleppestemmen house II	EBA I–II	Three-aisled group 2	W–E	ca 24	ca 9.0

Square metres	Roof-bear-ing posts	Distance between trestles	Wall posts	Entrances	Internal dividing walls	Fire places
87	3		WP	Opposing at east end	Eastern part	W part
106	12 (SF)		WP	Opposing at SW end?	Southern end?	Middle, towards SW
166	4		WP			
48	5 (SF	1–2.1	WP	Mid E end	Northern part?	
133	6	2.5–5.1 (RP)	WP			4, spread throughout
143	6	1–5 (RP)	WP	Mid E end		2, mid, along E wall
82	6		WP	E, close to inner wall	Northern part	S part close to E wall
65	3 (M) 1 (OC)	2–4	WP	Possible S wall	Middle	3, mid/N part
125	11 (5-6 PR)	2.4–2.7	PP	Opposing at middle		
178	10 (5 PR)	C. 3.5	WP	AP at S part		
49	6 (3 PR)	Varied	Wall ditch PP	1 entrance, NE corner		SE part
87	6 (3 PR)	3.4	Wall ditch PP			SE-part, middle
105	12 (5 PR)	2.7–3.75	Wall ditch PP		Possibly, just W of middle	5 (fireplaces and pits) in E part
127	14 (7 PR)	0.7–2.8	PP ditch along wall	Mid, NW-facing wall	Middle	3, N-part
90	12 (5–7 PR)	C. 2.5	P	O at middle		
114	8	c. 2.8	A few P	Two along W wall		NW part
104	11 (5–6 PR)	4.0–4.1		None seen		SW part
154	14 (7 PR)	Ca 3.5		AP at middle	Dividing walls in W and E (3 parts)	E and W
116	8 (4 PR)		P			
172	11 (5–6 PR)	3.5–3.7	P	Possibly AP at W	E End	
101	14 (7 PR)	2.2–2.6	P	O at middle		
167	16 (8 PR)	3.8–4.4	PP (wall ditch)	O at middle		
196	10 (5 PR)					NW part
216	12 (6 PR)	4.5–4.75				

Bibliography

Agamben, G. 2004. *The Open: Man and Animal*. Stanford, CA: Stanford University Press.

Albarella, U. 2001. "Exploring the real nature of environmental archaeology: An introduction." In *Environmental Archaeology: Meaning and Purpose*, edited by U. Albarella, 3–13. London: Kluwer Academic Publishers.

Albarella, U. and D. Serjeantson. 2002. "A passion for pork: Butchery and cooking at the British Neolithic site of Durrington Walls." In *Consuming Passions and Patterns of Consumption*, edited by N. Milner and P. Miracle, 33–49. Cambridge: McDonald Institute.

Alberti, B., S. Fowles, M. Holbraad, Y. Marshall and C. Whitmore. 2011. ""Worlds otherwise." Archaeology, anthropology and ontological difference." *Current Anthropology* 52(6): 896–912.

Allen, K., S. Barker, A. Beck, D. Bell, M. Frederickson, J. Freng, K. Gerbasi, E. Gordon, L. Hart, R. Johnson, J. Odendaal, A. O'Connor, M. Salman and B. Wong. 2007. "The health benefits of companion animals." http://www.nps.gov/goga/parkmgmt/upload/Comment-4704-attachment_.pdf

Almond, P., T. Edwards and I. Clark. 2003. "Multinationals and changing national business systems in Europe: Towards the 'shareholder value' model?" *Industrial Relations Journal* 34(5): 430–445.

Alrøe, H. F., M. Vaarst and E. S. Kristensen. 2001. "Does organic farming face distinctive livestock issues?: A conceptual analysis." *Journal of Agricultural and Environmental Ethics* 14(3): 275–299.

Appleby, D., J. W. S. Bradshaw and R. A. Casey. 2002. "Relationship between aggression and avoidance behaviour by dogs and their experience in their first six months of life." *The Veterinary Record* 150(14): 434–438.

Argent, G. 2010. "Do the clothes make the horse? Relationality, roles and statuses in Iron Age Inner Asia." *World Archaeology* 42(2): 157–174.

———. 2011. "At home, with the good horses: relationality, roles, identity and ideology in Iron Age Inner Asia." Unpublished PhD, University of Leicester.

———. 2012. "Toward a privileging of the nonverbal: Communication, corporeal synchrony and transcendence in humans and horses." In *Experiencing Animals: Encounters between Animal and Human Minds*, edited by J. A. Smith and R. W. Mitchell, 111–128. New York: Columbia University Press.

———. 2013. "Inked: Human-horse apprenticeship, tattoos, and time in the Pazyryk world." *Society & Animals* 21(2): 178–193.

Årlin, C. 1999. "Under samma tak - om 'husstallets' uppkomst och betydelse under bronsåldern ur ett sydskandinaviskt perspektiv." In *Spiralens öga. Tjugo artiklar kring aktuell bronsåldersforskning*, edited by M. Olausson, 291–307. Lund: Riksantikvarieämbetet.

Arluke, A. and C. R. Sanders. 1996. *Regarding Animals*. Animals, culture and society. Philadelphia, PA: Temple University Press.

Armstrong Oma, K. 2007. *Human-Animal relationships: Mutual becomings in the household of Scandinavia and Sicily 900-500 BC*. Oslo arkeologiske serie 9. Oslo: Unipub.

———. 2008a. "Kropp og sjel: Eit skeivt blikk på bronsealderdiskursen." *Nicolay* 104: 67–72.

———. 2008b. "Yoga in the Bronze Age? Reflections on pictorial evidence of the use of body techniques to attain altered states of consciousness." In *Facets of Archeology: Essays in Honour of Lotte Hedeager on her 60th birthday*, edited by K. Chilidis, J. Lund and C. Prescott, 485–496. Oslo: Unipub.

———. 2010. "Between trust and domination: Social contracts between humans and animals." *World Archaeology* 42(2): 175–187.

———. 2011. *Hesten: en magisk følgesvenn i nordisk forhistorie*. Oslo: Cappelen Damm.

———. 2012. "Large-scale "grand narratives" and small-scale local studies in the Bronze Age discourse: The animal perspective." In *Local Societies in Bronze Age Northern Europe*, edited by N. Anfinset and M. Wrigglesworth, 71–88. Sheffield: Equinox.

———. 2013a. "Bronze Age horses: Beyond dualist explanations." In *Counterpoint: Essays in Archaeology and Heritage Studies in Honour of Professor Kristian Kristiansen*, edited by S. Bergerbrant and S. Sabatini, 141–145. British Archaeological Reports International Series 2508. Oxford: Archaeopress.

———. 2013b. "Human–Animal meeting points: Use of space in the household arena in past societies." *Society & Animals* 21(2): 162–177.

———. 2013c. "Past and present farming: Changes in terms of engagement." In *Humans and the Environment: New Archaeological Perspectives for the 21st Century*, edited by M. I. Davies and F. Nkirote, 181–192. Oxford: Oxford University Press.

———. 2015. "Kor kjem me frå? Nye forteljingar om bronsealderen." *Fra Haug ok Heidni* 4: 15–17.

———. 2016a. "A shattered farm: Changes in making space from Pagan to Christian Norway." In *The Farm As A Social Arena*, edited by L. H. Dommasnes, A. T. Hommedal and D. Gütschmidl, 171–190. Münster: Waxmann Verlag.

———. 2016b "From horses to Jesus—saving souls in the transition from pagan to Christian Scandinavia." In *The Meaning of Horses: Biosocial Encounters*, edited by D. Davis and A. Maurstad, 24–38. London and New York: Routledge.

———. 2016c. "Long time—long house: Agrarian commitments, households and ontology from the Early Bronze Age to the Viking period in Rogaland, Norway." In *Agrarian Living in the North BC 1800-1000 AD: Studies in Rural Settlements and Farming in Norway*, edited by F. Iversen and H. Petersson, 11–25. Kristiansand: Portal forlag.

———. 2018. "Making space from the position of duty of care: Early Bronze Age human-sheep entanglements in Norway." In *Multispecies Archaeology*, edited by S. E. Pilaar Birch, 214–229. London: Routledge.

Armstrong Oma, K. and L. Birke. 2013. "Archaeology and Human–Animal Studies." *Society & Animals* 21(2): 113–119.

Armstrong Oma, K. and A. L. Melheim. In prep. "'Children of the light.' On yoga, body schemes and altered states of consciousness in the Nordic Late Bronze Age—a link to India?" In *Tracing the Indo-Europeans*, edited by B. S. Sandgaard Hansen, K. Kristiansen, G. Kroonen, T. Olander, B. A. Rasmussen Olsen and O. B. Simkin.

Arthursson, M. 2009. "Bebyggelse och samhällsstruktur: södra och mellersta Skandinavien under senneolitikum och bronsålder 2300–500 f. Kr." In *GOTARC Serie B*. Gothenburg Archaeological Thesis 52/Riksantikvarieämbetet, Arkeologiska undersökningar. Skrifter 73. Göteborg: Göteborg University/Stockholm.

Austvoll, K. I. 2014. "Costructing identities: Structure and practice in the Early Bronze Age—Southwest Norway." Unpublished MA thesis, University of Oslo.

Barad, K. 2007. *Meeting the Universe Halfway. Quantum Physics and the Entanglement of Matter and Meaning*. Durham, NC: Duke University Press.

Barham, A. J. and R. I. Macphall, eds. 2009. *Archaeological Sediments and Soils: Analysis, Interpretation and Management*. UCL Institute of Archaeology Publications 4. London: Routledge.

Barker, G. 1999. "Cattle-keeping in ancient Europe: To live together or apart?" In *Settlement and Landscape*, edited by C. Fabeck and J. Ringtved, 273–280. Århus: Jutland Archaeological Society.

Barrett, J. C. 2000. "Fields of discourse: Reconstituting a social archaeology." In *Interpretive Archaeology: A Reader*, edited by J. Thomas, 23–32. London: Leicester University Press.

Bear, C., L. Holloway and K. Wilkinson. 2012. "Robotic milking technologies and the renegotiation of situated ethical relationships on UK dairy farms." Paper presented at the British Animal Studies Network Second Meeting, FARM, Glasgow, 17.11.2012.

Bech, J.-H. 1991. "Et bronzealderlandskab ved Bjerre i Nordthy. Om arkæologiske udgravninger forud for en planlagt motorbane." In *Museerne i Vibord Amt Aarbog* 16: 41–48.

———. 1997. "Bronze Age settlements on raised sea-beds at Bjerre, Thy, NW-Jutland." In *Forschungen zur bronzezeitlichen Besiedlung in Nord- und Mitteleuropa. Internationales symposium von 9-11 Mai 1996 in Hitzacher*, edited by J. J. Assendorp, 3–15. Espelkamp: Internationale Archäelogie.

Bech, J.-H. and M. Mikkelsen. 1999. "Landscape, settlement and subsistence in Bronze Age Thy, NW Denmark." In *Settlement and Landscape: Proceedings of a Conference in Århus, 1998*, edited by C. Fabech and J. Ringtved, 69–78. Moesgård: Jutland Archaeological Society.

Bech, J.-H. and A.-L. H. Olsen. 2013. "Early Bronze Age houses from Thy, Northwest Denmark." In *Studien zur nordeuropäischen Bronzezeit* edited by K.-H. Willroth, 9–32. Neumünster: Wachholtz Verlag.

Behre, K. E. 1998. "Landwirschaftliche Entwiklungslinien und die Veränderung der Kulturlandschaft in der Bronzezeit Europas." In *Mensch und Umwelt in der Bronzezeit Europas*, edited by B. Hänsel, 91–109. Kiel: Oetker-Voges Verlag.

Bekoff, M. 2002. *Minding Animals: Awareness, Emotion and Heart*. Oxford: Oxford University Press.

———. 2007. *The Emotional Lives of Animals*. Novato, CA: New World Library.

———. 2012a. "Dogs, dominance and Cesar Millan redux: Dominance is real." *Psychology Today*. http://www.psychologytoday.com/blog/animal-emotions/201212/dogs-dominance-and-cesar-millan-redux-dominance-is-real.

———. 2012b. "Scientists finally conclude nonhuman animals are conscious beings." *Psychology Today*. http://www.psychologytoday.com/blog/animal-emotions/201208/scientists-finally-conclude-nonhuman-animals-are-conscious-beings.

Benecke, N. 1994. *Archäozoologische Studien zur Entwicklung der Haustierhaltung in Mitteleuropa und Südskandinavien von der Anfängen bis zum ausgehenden Mittelalter*. Vol. 46, *Schriften zur Ur- und Frühgeschichte*. Berlin: Akademie Verlag.

Bellows, Henry Adams, ed. and trans. 1936a. "Fragment of a Sigurd Lay." In *The Poetic Edda*. http://www.sacred-texts.com/neu/poe/poe26.htm

———. 1936b. "The second lay of Gunhild." In *The Poetic Edda*. http://www.sacred-texts.com/neu/poe/poe31.htm

Berge, J. 2007. "Undersøking av hustomter frå slutten av yngre steinalder og eldre bronsealder, Kleppe, gnr. 1, bnr. 22, Kleppestemmen, Klepp kommune. Oppdragsrapport B." Stavanger: Arkeologisk museum i Stavanger.

Berger, J. 1971. "Animal world." *New Society* 18: 1042–1043.

———. 2009 [1977]. *Why Look at Animals?* London: Penguin Books.

Bergerbrant, S. 2007. *Bronze Age Identities: Costume, Conflict and Contact in Northern Europe 1600-1300 BC*. Stockholm Studies in Archaelogy 43. Lindome: Bricoleur Press.

———. 2014. "Ordinary or extraordinary? Redressing the problem of the Bronze Age corded skirt." *Current Swedish Archaeology* 22: 73–96.

Bertilsson, U. 1994. "Hällristningar och bygden." In *Hällristningar och hällmålningar i Sverige*, edited by U. Bertilsson, S. Jansson and E. B. Lundberg, 29–42, 82–120. Stockholm: Forum.

Binford, L. R. 1978. *Nunamiut Ethnoarchaeology*. New York: Academic Press.

———. 1981. *Bones: Ancient Men and Modern Myths*. New York: Academic.

Birke, L. 2002. "Intimate familiarities? Feminism and Human-Animal Studies." *Society and Animals* 10: 429–436.

———. 2010. "Structuring relationships: On science, feminism and non-human animals "*Feminism & Psychology* 20(3): 337–349.

———. 2012. "Unnamed others: How can thinking about 'animals' matter to feminist theorizing?" *NORA—Nordic Journal of Feminist and Gender* 20(2): 148–157.

Birke, L., M. Bryld and N. Lykke. 2004. "Animal performances": An exploration of intersections between feminist science studies and studies of human/animal relationship." *Feminist Theory* 5(2): 167–183.

Birke, L. and J. Hockenhull. 2012a. "Introduction: On investigating human-animal relationships." In *Crossing Boundaries: Investigating Human-Animal Relationships*, edited by L. Birke and J. Hockenhull, 1–12. Leiden: Brill.

———. 2012b. "On investigating human-animal bonds: Realities, relatings, research." In *Crossing Boundaries: Investigating Human-Animal Relationships*, edited by L. Birke and J. Hockenhull, 15–36. Leiden: Brill.

Birke, L. and L. Parisi. 1999. "Animals, becoming." In *Animal Others: On Ethics, Ontology and Animal Life*, edited by H. P. Steeves, 55–73. Albany: State University of New York Press.

Bjørlo, A. 2011. "Arkeologisk og naturvitensakpelig undersøkelse av to-skipet langhus fra eldre bronsealder og branngrav fra eldre jernalder på Kvia. Kvia gnr. 19, bnr. 31, Motland gnr. 20, bnr. 3, Hå kommune, Rogaland." *Oppdragsrapport* 2011/28. Stavanger: Universitetet i Stavanger, Arkeologisk museum.

Bökönyi, S. 1975. "Effects of environmental and cultural changes on prehistoric fauna assemblages." In *Gastronomy: The anthropology of Food and Food Habits*, edited by M. L. Arnott, 3–12. The Hague: Mouton Publishers.

Borna-Ahlkvist, H. 2002. *Hällristarnas hem. Gårdsbebyggelse och struktur i Pryssgården under bronsålder.* Riksantikvarieämbetet arkeologiske undersökningar skrifter 42. Malmö: Riksantikvarieämbetet.

Borna-Ahlkvist, H., L. Lindgren-Hertz and U. Stålbom. 1998. *Pryssgården. Från stenålder till medeltid. Arkeologisk slutundersökning RAÄ 166 och 167 Östra Eneby socken Norrköpings kommun Östergötland, vol. 13, Rapport UV*. Linköping: Riksantikvarieämbetet. Avdelningen för arkeologiska undersökningar.

Børresen, B. 2004. *Den store boken om husdyrene og menneskene deres*. Oslo: Gyldendal.

Bourdieu, P. 1977. *Outline of a Theory of Practice*. Cambridge: Cambridge University Press.

———. 1990 [1980]. *The Logic of Practice*. Stanford, CA: Stanford University Press.

Bradley, R. 2005. *Ritual and Domestic Life in Prehistoric Europe*. London: Routledge.

Bradshaw, J. 2011. *In Defense of Dogs*. London: Allen Lane.

Brittain, M. and N. Overton. 2013. "The significance of others: A prehistory of rhythm and interspecies participation." *Society & Animals* 21(2): 134–149.

Budiansky, S. 1992. *The Covenant of the Wild: Why Animals Chose Domestication*. New York: William Morrow.

———. 1994. "A special relationship: The coevolution of human beings and domesticated animals." *Journal of the American Veterinary Medical Association* 204(3): 365–368.

Buller, H. 2012. "'One slash of light, then gone.' Animals as movement." *Études Rurales* 189(1): 139–154.

Butler, J. 1993. *Bodies That Matter: On the Discursive Limits of "Sex."* London: Routledge.

Callicott, J. B. 1989. *In Defense of the Land Ethic*. New York: Albany.

Carlie, L. 2002. "Järnålderns mångfunktionella långhus, myt eller verklighet." *In Situ Archaeologica* 4: 61–72.

Cassidy, R. 2007. "Introduction: domestication reconsidered." In *Where the Wild Things Are Now: Domestication Reconsidered*, edited by R. Cassidy and M. H. Mullin, 1–25. New York: Berg.

Chapman, J. 1983. "The secondary products revolution and the limitations of the Neolithic." *University of London Institute of Archaeology Bulletin* 19: 107–122.

Chessa, B., F. Pereira, F. Arnaud, A. Amorim, F. Goyache, I. Mainland, Rowland R. Kao1, Josephine M. Pemberton6, D. Beraldi, M. J. Stear, A. Alberti, M. Pittau, L. Iannuzzi, M. H. Banabazi, R. R. Kazwala, Y.-p. Zhang, J. J. Arranz, B. A. Ali, Z. Wang, M. Uzun, M. M. Dione, I. Olsaker, L.-E. Holm, U. Saarma, S. Ahmad, N. Marzanov, E. Eythorsdottir, M. J. Holland, P. Ajmone-Marsan, M. W. Bruford, J. Kantanen, T. E. Spencer and M. Palmarini. 2009. "Revealing the history of sheep domestication using retrovirus integrations." *Science* 324(5926): 532–536.

Clutton-Brock, J., ed. 1989. *The Walking Larder: Patterns of Domestication, Pastoralism and Predation*. One World Archaeology. London: Routledge.

———. 1999 [1987]. *A Natural History of Domesticated Mammals*. Cambridge: Cambridge University Press.

Colledge, S., J. Connolly and S. Shennan. 2005. "The evolution of early Neolithic agriculture: From SW Asian origins to NW European limits." *European Journal of Archaeology* 8(2): 137–156.

Conneller, C. 2004. "Becoming deer. Corporeal transformations at Star Carr." *Archaeological Dialogues* 11(1): 37–56.

Coren, S. 1995. *The Intelligence of Dogs: A Guide to the Thoughts, Emotions and Inner Lives of Our Canine Companions*. New York: Bantam Books.

Cudworth, E. 2011. *Social Lives with Other Animals: Tales of Sex, Death and Love*. Basingstoke: Palgrave Macmillan.

Dabbert, S., A. M. Häring and R. Zanoli. 2004. *Organic Farming: Policies and Prospects*. London: Zed Books.

Dahl, B. I. 2009. "En presentasjon av fire utvalgte hus fra Forsandmoen 2007." In *Tverrfaglige perspektiver*, edited by M. Nitter and E. Solhem Pedersen, 87–104. Stavanger: Arkeologisk museum, Universitetet i Stavanger.

———. 2014. "Arkeologisk utgraving av hus og graver. Myklebust gnr 3, Sola kommune, Rogaland." *Oppdragsrapport* 2014/20. Stavanger: Arkeologisk museum, Universitetet i Stavanger.

Dallery, C. 1999. "Into the truth with animals." In *Animal Others: On Ethics, Ontology and Animal Life*, edited by H. P. Steeves 249–270. Albany: State University of New York Press.

Darwin, C. 1936 [1871]. *The Descent of Man and Selection in Relation to Sex*. New York: Random House.

Dawkins, M. S. 1980. *Animal suffering: The Science of Animal Welfare*. London: Chapman & Hall.

Delanda, M. 2005. *Intensive Science and Virtual Philosophy*. London: Continuum.

Deleuze, G. and F. Guattari. 2004 [1987]. *A Thousand Plateaus: Capitalism and Schizophrenia*. New York: Continuum.

DeMarrais, E., C. Gosden and C. Renfrew, eds. 2004. *Rethinking Materiality: the Engagement of Mind with the Material World*. MacDonald Institute Monographs. Cambridge: MacDonald Institute for Archaeological Research.

DeMello, M. 2012. *Animals and Society: An Introduction to Human-Animal Studies*. New York: Columbia University Press.

Denenberg, V. H. 1962. "The effects of early experience." In *The Behaviour of Domestic Animals*, edited by E.S.E. Hafez, 109–138. London: Baillière, Tindall & Cox.

Denham, T. and P. White. 2007. *The Emergence of Agriculture: A Global View*. One World Archaeology Readers 1. London: Routledge.

Derrida, J. 2008. *The Animal That Therefore I Am*. New York: Fordham University Press.

Descartes, R. 1912 [1637]. *Discourse on Method*. New York: E. P. Dutton.

Dietler, M. and B. Hayden. 2001. "Digesting the feast—good to eat, good to drink, good to think: An introduction." In *Feasts: Archeological and Ethnographic Perspectives on Food, Politics and Power*, edited by M. Dietler, B. Hayden, 1–20. Washington, DC: Smithsonian Institution Press.

Donaldson, T. and L. E. Preston. 1995. "The stakeholder theory of the corporation: Concepts, evidence and implications." *The Academy of Management Review* 20(1): 65–91.

Douglas, M. 1966. *Purity and Danger: An Analysis of Concepts of Pollution and Taboo*. London: Routledge & Kegan Paul.

Dransart, P. Z. 2002. *Earth, Water, Fleece and Fabric: An Ethnography and Archaeology of Andean Camelid Herding*. New York: Routledge.

Earle, T. 2002. *Bronze Age Economics: The Beginnings of Political Economics*. Cambridge, MA: Westview Press.

Edelman, B. 2005. "From trap to lap: The changing sociogenic identity of the rat." In *Animals in Person: Cultural Perspectives on Human-Animal Intimacy*, edited by J. Knight, 119–139. New York: Berg.

Eilertsen, K. S. In prep 2018. Tanangervegen: Arkeologisk undersøkelse av boplasser fra eldre- og yngre steinalder, samt bosetningsspor fra yngre steinalder, bronse- og eldre jernalder. Sømme gnr. 15, Sola kommune. Rogaland. *Oppdragsrapport* 2018/5. Stavanger: Arkeologisk museum i Stavanger.

Engedal, Ø. 2010. "The Bronze Age of Northwestern Scandinavia." Unpublished PhD thesis, University of Bergen.

Engelstad, E. 2001. "Gender, feminism, and sexuality in archaeological studies." In *International Encyclopedia of the Social and Behavioral Sciences*, edited by N. J. Smelser and P. B. Baltes, 6002–6006. Amsterdam: Elsevier.

Estevez, I., I. L. Andersen and E. Naevdal. 2007. "Group size, density and social dynamics in farm animals." *Applied Animal Behaviour Science* 103(3-4): 185–204.

Ethelberg, P. 1995. "Nr. 579, Brd. Gram. Arkæologiske udrgavninger i Danmark 1992." Unpublished report.

Ethelberg, P., L. S. Madsen and O. Madsen. 2000. *Det Sønderjyske landbrugs historie*. Vol. 81, *Skrifter udgivet af Historisk Samfund for Sønderjylland*. Haderslev: Haderslev Museum og Historisk Samfund for Sønderjylland.

Festa-Bianchet, M. 1991. "The social system of sheep: Grouping patterns, kinship and female dominance rank." *Animal Behaviour* 42(1): 71–82.

Fett, E. and P. Fett. 1941. *Sydvestnorske helleristninger. Rogaland og Lista*. Stavanger: Dreyer.

Fijn, N. 2011. *Living with Herds: Human-Animal Coexistence in Mongolia*. Cambridge: Cambridge University Press.

Fisher, A. and L. Matthews. 2001. "The social behaviour of sheep." In *Social behaviour in Farm Animals*, edited by H. W. Gonyou and L. J. Keeling, 211–245. Wallingford, Oxon: The Centre for Agriculture and Bioscience International.

Fokkens, H. 1991. "Nederzettingssporen uit de bronstijd en de vroge ijzertijd in Oss-Ussen, wijk Middeldonk." In *Nederzettingen uit de bronstijd en de vroege ijzertijd in de Lage Landen*, edited by H. Fokkens and N. Roymans, 93–109. Amersfoort: Rijksdienst voor het Oudheidkundig Bodemonderzoek.

———. 1998. "From the collective to the individual: Some thoughts about social and economic change around 3000 BC." In *Understanding the Neolithic of North-Western Europe*, edited by M. R. Edmonds and C. Richards, 481–491. Glasgow: Cruithne Press.

———. 1999. "Cattle and martiality: Changing relations between man and landscape in the Late Neolithic and the Bronze Age." In *Settlement and Landscape*, edited by C. Fabeck and J. Ringtved, 35–43. Århus: Jutland Archaeological Society.

Forrest, C., L. Goldman and M. Emmison. 2005. "The unbearable likeness of being: Children, teddy bears and *The Sooty Show*." In *Animals in Person: Cultural Perspectives on Human-Animal Intimacy*, edited by J. Knight, 141–161. New York: Berg.

Franklin, A. 1999. *Animals and Modern Culture: A Sociology of Human-Animal Relations in Modernity*. London: Sage.

———. 2002. *Nature and Social Theory*. London: Sage Publications.

Fredengren, C. 2013. "Posthumanism, the transcorporeal and biomolecular archaeology." *Current Swedish Archaeology* 21: 53–71.

Fredsjö, Å. 1969. "Hällristningar i västra Sverige." In *Hällristningar i Sverige*, edited by Å. Fredsjö, S. Janson and C.-A. Moberg, 95–134. Stockholm: Forum.

Freeman, R. E. 1983. *Strategic Management: A Stakeholder Approach*. Pitman Series in Business and Public Policy. Boston, MA: Financial Times Prentice Hall.

Frei, K. M., U. Mannering, K. Kristiansen, M. E. Allentoft, A. S. Wilson, I. Skals, S. Tridico, M.-L. Nosch, E. Willerslev, L. Clarke and R. Frei. 2015. "Tracing the dynamic life story of a Bronze Age female." *Nature Scientific Reports* 5 (10431): 1–7. http://www.nature.com/articles/srep10431.pdf

Frei, K. M., U. Mannering, I. Vanden Berghe and K. Kristiansen. 2017a. "Bronze Age wool: Provenance and dye investigations of Danish textiles." *Antiquity* 91(357): 640–654.

Frei, K. M., C. Villa, M. L. Jørkov, M. E. Allentoft, F. Kaul, P. Ethelberg, S. S. Reiter, A. S. Wilson, M. Taube, J. Olsen, N. Lynnerup, E. Willerslev, K. Kristiansen and R. Frei. 2017b. "A matter of months: High precision migration chronology of a Bronze Age female." *PLoS ONE* 12 (6):e0178834. https://doi.org/10.1371/journal.pone.0178834.

Fudge, E. 2008. *Pets*. Stocksfield: Acumen.

Fudge, E., R. Gilbert and S. Wiseman, eds. 1999. *At the Borders of the Human: Beasts, Bodies and Natural Philosophy in the Early Modern Period*. Basingstoke: Palgrave.

Fyllingen, H. 2012. "Arkeologisk utgravning på Jåsund gnr.1, bnr. 6, 10, 19/20 & 27. Sola kommune Rogaland." *Oppdragsrapport* 2012/22. Stavanger: Universitetet i Stavanger, Arkeologisk museum.

———. 2015a. "Arkeologiske undersøkelser på gårdene Bjorhaug gnr. 25 og Bø/Bøhagen gnr. 26 i Hå kommune. Et bidrag til innblikk i gårdsstrukturen på Jæren i perioden 1500 f.Kr. til 800 e.Kr." *Oppdragsrapport* 2015/4. Stavanger: Arkeologisk museum, Universitetet i Stavanger.

———. 2015b. *Resultater fra undersøkelsene av et gravfelt, en flyttblokk og bosetningsspor på Tjora, gnr. 10, Sola kommune, Rogaland.* Vol. 2, *AM-Profil.* Stavanger: Arkeologisk museum, Universitetet i Stavanger.

Fyllingen, H. and N. J. O. Armstrong. 2012. Arkeologisk utgraving på Tjora, gnr 10 bnr 5, 17, og 19. Sola kommuna, Rogaland. Sesong 2009 - id.114908. *Oppdragsrapport* 2017/23B. Stavanger: Arkeologisk museum, Universitetet i Stavanger.

Gaillard, M.-J., H. J. B. Birks, U. Emanuelson, S. Karlsson, P. Lagerås and D. Olausson. 1994. "Application of modern pollen/land-use relationships to the interpretation of pollendiagrams—reconstructions of land-use history in south Sweden 3000-0 BP." *Review for Paleobotany and Palynology* 82: 47–73.

Game, A. 2001. "Riding: embodying the Centaur." *Body & Society* 7(4): 1–12.

Gero, J. and W. Conkey, eds. 1991. *Engendering Archaeology.* Oxford: Blackwell.

Gilchrist, R. 1999. *Gender and Archaeology: Contesting the Past.* London: Routledge.

Gittins, E. 2013. "The archaeology of becoming the human animal." *Society & Animals* 21(2): 120–133.

Glob, P. V. 1969. *Helleristninger i Danmark.* Jysk Arkæologisk Selskabs Skrifter vol. VII. Copenhagen: Jutland Archaeological Society.

Goldhahn, J. 2006. *Hällbildsstudier i norra Europa - trender och tradition under det nya millenniet.* Gothenburg: Gotarc serie C, Arkeologiska skrifter.

Goldhahn, J., I. Fuglestvedt and A. Jones, eds. 2010. *Changing Pictures: Rock Art Traditions and Visions in Northernmost Europe.* Oxford: Oxbow Books.

Goodwin, B. 1999. "Reclaiming a life of quality." *Journal of Consciousness Studies* 6: 229–235.

Grandin, T. 1987. "Animal handling." In *Farm animal behaviour*, edited by E. O. Price, 323-338. Philadelphia: Saunders.

Grant, A. 2002. "Food, status and social hierarchy." In *Consuming passions and patterns of consumption*, edited by N. Milner, P. Miracle, 17-23. Cambridge: McDonald Institute.

Gräslund, A.-S. 2002. "Dogs in graves - a question of symbolism?" PECUS. Man and animal in antiquity, Swedish Institute in Rome.

Greenfield, H. J. 2005. "A reconsideration of the Secondary Products Revolution in south-eastern Europe: on the origins and use of domestic animals for milk, wool and traction in the central Balkans." In *The zooarchaeology of fats, oils, milk and dairying. Proceedings from the 9th ICAZ conference, Durham 2002*, edited by J. Mulville and A. K. Outram, 14-31. Oxford: Oxbow.

Grew, S. 1993. *Key dogs from the Border Collie family.* Hernando, MS: Heritage Farms Publishers.

Grimstvedt, M., T. Hertel-Aas and L. G. Jøssang. 1996. *Med sauen til heis, Jærskrifter 2.* Nærbø: Jærmuseet.

Grön, A. 2004. *Positioning the Bronze Age in Social Theory and Research Context.* Acta Archaeologica Lundensia Series in 8°, vol. 47. Lund: Almqvist & Wisell International.

Hallowell, I. 2010. *Contributions to Ojibwe Studies: Essays, 1934-1972.* Edited by J.S.B. Brown and S. E. Gray. Lincoln: University of Nebraska Press.

Halsall, E. 1985. *Sheepdogs: My faithful Friends.* Wellingborough: Patrick Stephens.

Halstead, P. 1998. "Mortality models and milking: Problems of uniformitarianism, optimality and equifinality reconsidered." *Anthropozoologica* 27: 3–20.

Halstead, P. and V. Isaakidou. 2004. "Faunal evidence for feasting: Burnt offerings from the palace of Nestor at Pylos." In *Food, Cuisine and Society in Prehistoric Greece*, edited by P. Halstead, J. C. Barrett, 136–154. Oxford: Oxbow Books.

Halstead, P. and J. C. Barrett, ed. 2004. *Food, Cuisine and Society in Prehistoric Greece.* Sheffield Studies in Aegean Archaeology, vol. 5. Oxford: Oxbow Books.

Hamilakis, Y. 1999. "Food technologies/technologies of the body: The social context of wine and oil production and consumption in Bronze Age Crete." *World Archaeology* 31(1): 38–54.

———. 2001. "Re-inventing environmental archaeology? A comment on 'Economic prehistory or environmental archaeology? On gaining a sense of identity'." In *Environmental Archaeology: Meaning and Purpose*, edited by U. Albarella, 29–38. London: Kluwer Academic Publishers.

———. 2003. "The sacred geography of hunting: Wild animals, social power and gender in early farming societies." In *Zooarchaeology in Greece: Recent Advances*, edited by E. Kotjabopoulou, 239–247. London: British School at Athens.

Hamilakis, Y. and E. Konsolaki. 2004. "Pigs for the Gods: Burnt animal sacrifices at a Mycenaean sanctuary." *Oxford Journal of Archaeology* 23(2): 135–151.

Handlin, L. 2010. "Human-Human and Human–Animal Interaction: Some Common physiological and psychological effects." Unpublished PhD, Faculty of Veterinary Medicine and Animal Science, Department of Animal Environment and Health, Swedish University of Agricultural Sciences.

Haraway, D. 1988. "Situated knowledges: The science question in feminism and the privilege of partial perspectives." *Feminist Studies* 14(3): 575–599.

———. 2003. *The Companion Species Manifesto: Dogs, People and Significant Otherness*. Chicago, IL: Prickly Paradigm Press.

———. 2004. "Cyborgs to companion species: Reconfiguring kinship in technoscience." In *The Haraway Reader*, 295–320. London: Routledge.

———. 2008. *When Species Meet*. Posthumanities, vol. 3. Minneapolis: University of Minnesota Press.

Harding, A. F. 2000. *European Societies in the Bronze Age*. Cambridge World Archaeology. Cambridge: Cambridge University Press.

Harding, S. 1986. *The Science Question In Feminism*. London: Open University.

Harsema, O. H. 1993. "Het nieuwe Drentse bouwbesluit van 1400 v.Chr." *Westerheem* 42(3): 101–109.

Hatkoff, A. 2009. *The Inner World of Farm Animals: Their Amazing Social, Emotional and Intellectual Capacities*. New York: Stewart, Tabori and Chang.

Hayden, B. 1996. "Feasting in prehistoric and traditional societies." In *Food and the Status Quest: An Interdisciplinary Perspective*, edited by P. W. S. Wiessner, 127–147. Providence, RI: Berghahn Books.

Hedeager, L. 1999. "Skandinavisk dyreornamentik. Symbolsk repræsentasjon af en førkristen kosmologi." In *Et hus med mange rom. Vennebok til Bjørn Myhre på 60-årsdagen*, edited by I. Fuglestvedt, T. Gansum, A. Opedal, 219–238. Stavanger: Arkeologisk museum i Stavanger.

———. 2011. *Iron Age Myth and Materiality: An Archaeology of Scandinavia AD 400–1000*. London: Routledge.

Heidegger, M. 1962. *Being and Time*. London: SCM Press.

Heinrichs, M. and G. Domes. 2008. "Neuropeptides and social behaviour: Effects of oxytocin and vasopressin in humans." In *Advances in Vasopressins and Oxytocin: From Genes to Behaviour to Disease*, edited by I. D. Neumann and R. Landgraf, 337-350. Oxford: Elsevier.

Helms, M. W. 1988. *Ulysses' Sail: An Ethnographic Odyssey of Power, Knowledge and Geographical Distance*. Princeton, NJ: Princeton University Press.

———. 1993. *Craft and the Kingly Ideal: Art, Trade and Power*. Austin: University of Texas Press.

———. 1998. *Access to Origins: Affines, Ancestors and Aristocrats*. Austin: Texas University Press.

Hemdorff, O. 1989. "Innberetning til topografisk arkiv. Innberetning om arkeologisk undersøkelse af lokalitet 20 på Austbø gnr. 7, bnr. 7, Hundvåg, Stavanger kommune." Unpublished report. Stavanger: Arkeologisk museum i Stavanger.

———. 1993a. "Hus fra eldre bronsealder funnet på Talgje." *Frá haug ok heidni* 4: 23–26.

———. 1993b. "Innberetning om de arkeologiske undersøkelser på Østabø gnr. 43. Bnr. 1 Talgje, Finnøy kommune. Innberetning til topografisk arkiv." Unpublished report. Stavanger: Arkeologisk museum i Stavanger.

Hernes, P. 1992. "Nytt funn av helleristninger på Ølbør i Sola." *Fra haug ok heidni* 4: 7–8.

Hiendleder, S., B. Kaupe, R. Wassmuth and A. Janke. 2002. "Molecular analysis of wild and domestic sheep questions current nomenclature and provides evidence for domestication from two different subspecies." *Proceedings of the Royal Society: Biological Sciences* 269(1494): 893–904. http://rspb.royalsocietypublishing.org/content/269/1494/893

Higgs, E. S. 1972. *Papers in Economic Prehistory*. Cambridge: Cambridge University Press.

Hill, E. 2013. "Archaeology and animal persons: Toward a prehistory of human-animal relations." *Environment & Society: Advances in Research* 4: 117–136.

Hill, J. D. 1996. "The identification of ritual deposits of animal bones. A general perspective from a specific study of "special animal deposits" from the Southern English Iron Age." In *Ritual Treatment of Human and Animal Bones*, edited by S. Anderson and K. Boyle, 17–32. Oxford: Oxbow.

Hodder, I. 1992. *Theory and Practice in Archaeology*. London: Routledge.

———. 2014. *Çatalhöyük Excavations: The 2000–2008 Seasons*. Çatalhöyük Reseach Project Series Volume 7, British Institute at Ankara Monograph No. 46. Monumenta Archaeologica 29. Los Angeles: Cotsen Institute of Archaeology Press.

Høgestøl, M. and L. Prøsch-Danielsen. 2006. "Impulses of agro-pastoralism in the 4th and 3rd millennia BC on the south-western coastal rim of Norway." *Environmental Archaeology* 11(1): 19–34.

Holst, M. K. and M. Rasmussen. 2013. "Herder communities: Longhouses, cattle and landscape organization in the Nordic early and middle Bronze Age." In *Counterpoint: Essays in Archaeology and Heritage Studies in Honour of Professor Kristian Kristiansen*, edited by S. Bergerbrant and S. Sabatini, 99–110. British Archaeological Reports International Series 2508. Oxford: Archaeopress.

Hornstrup, K. M. 2011. "Rogaland - Thy i ældre bronsealder." In *Historisk årbog for Thy og Vester Hanherred*, 64–76. Thisted: Historisk Samfund for Thy og Vester Hanherred.

Horve, I. 2009. "Gård og samfunn. Økonomi og sosial struktur med utgangspunkt i bronsealderens langhus i Rogaland, Sørvest-Norge." Unpublished MA Thesis, University of Bergen.

Hosmer, L. T. 1995. "Trust: The connecting link between organizational theory and philosophical ethics." *The Academy of Management Review* 20(2): 379–403.

Hufthammer, A. K. 1994. "Tidlig husdyrhold i Vest-Norge." *Arkeologiske Skrifter fra Arkeologisk Institutt, Bergen Museum* 8: 203–219.

———. 2002. "Analyse av beinmaterialet fra Kvåle, Time kommune, Rogaland." Rapport fra Zoologisk museum, Universitetet i Bergen, i Topografisk arkiv, Arkeologisk museum i Stavanger. Stavanger: University of Stavanger.

———. 2004. "JS081 Rapport Forsandmoen, Forsand K., Rogaland." Arkeologisk Museum i Stavanger. Stavanger: University of Stavanger.

Hutson, G. D. 1985. "The influence of barley food rewards on sheep movement through a handling system." *Applied Animal Behaviour Science* 14(3): 263–284.

Hutson, W. H. 1993 [1910]. *The illustrated Shepherd's Life*. London: Bracken Books.

Iacoboni, M. 2009. *Mirroring People: The Science of Empathy and How We Connect with Others*. New York: Picador.

IJzereef, G. 1981. *Bronze Age Animal Bones from Bovenkarspel: The Excavation at Het Valkje*. Nederlandse Oudheden 10. Amersfoort: ROB.

Ingold, T. 2000a. "From trust to domination: An alternative history of human-animal relations." In *The Perception of the Environment: Essays in Livelihood, Dwelling and Skill*, edited by T. Ingold, 61–76. London: Routledge.

———. 2000b. *The Perception of the Environment: Essays in Livelihood, Dwelling and Skill.* London: Routledge.

———. 2011. *Being Alive: Essays on Movement, Knowledge and Description.* London: Routledge.

Irvine, L. 2004. *If You Tame Me: Understanding Our Connection With Animals.* Philadelphia, PA: Temple University Press.

Jamieson, D. and M. Bekoff. 1993. "On aims and methods of cognitive ethology." *Philosophy of Science Association* 2: 110–124.

Johannsen, N. 2012. "Archaeology and the Inanimate Agency Proposition: a critique and a suggestion." In *Excavating the Mind: Cross-sections through Culture, Cognition and Materiality*, edited by N. Johannsen, M. Jessen and H. Juel Jensen, 305–347. Aarhus: Aarhus University Press.

Jones, A. and C. Richards. 2003. "Animals into ancestors: Domestication, food and identity in Late Neolithic Orkney." In *Food, Culture and Identity in the Neolithic and Early Bronze Age*, edited by M. Parker Pearson, 45–52. British Archaeological Reports International Series 1117. Oxford: Archaeopress.

Jones, M. 2002. "Eating for calories or for company? Concluding remarks on consuming passions." In *Consuming Passions and Patterns of Consumption*, edited by N. Milner and P. Miracle, 131–136. Cambridge: McDonald Institute.

Jones, P. L. 2009. "Considering living-being in the aceramic Neolithic of Cyprus." *Journal of Mediterranean Archaeology* 22: 75–99.

Jones, T. M. 1995. "Instrumental stakeholder theory: A synthesis of ethics and economics." *The Academy of Management Review* 20(2): 404–437.

Juhl, K. 2001. *Austbø på Hundvåg gennem 10 000 år. Arkeologiske undersøgelser i Stavanger kommune 1987-1990 Rogaland, Syd-Vest Norge.* AmS-Varia 38. Stavanger: Arkeologisk museum i Stavanger.

Kaliff, A. 1998. "Grave structures and altars: Archaeological traces of Bronze Age eschatological conceptions." *European Journal of Archaeology* 1(2): 177–198.

———. 2005. "Den vediske *Agni* och skandinaviska eldritualer - ett möjligt samband?" In *Mellan sten och järn. Rapport från det 9:de nordiska bronsålders-symposiet, Göteborg 2003-10-09/12*, edited by J. Goldhahn, 167–189. Arkeologiska Skrifter 65. Göteborg: Gotarc Serie C.

Karlenby, L. 1994. "The Bronze Age house in central Sweden: An evaluation of two recent excavations." *TOR* 26: 5–33.

Kaul, F. 1998. *Ships on Bronzes: A Study in Bronze Age Religion and Iconography.* Studies in Archaeology and History 3. Copenhagen: Publications from the National Museum.

———. 2004. *Bronzealderens Religion: Studier af den Nordiske Bronzealders Ikonografi.* Nordiske Fortidsminder Vol. Serie B, Bind 22. København: Det Kongelige Nordiske Oldskriftselskab.

Kendrick, K. M. 1998. "Intelligent perception." *Applied Animal Behaviour Science* 57(3–4): 213–231.

Kendrick, K. M., A. P. da Costa, A. E. Leigh, M. R. Hinton and J. W. Peirce. 2001. "Sheep don't forget a face." *Nature* 414: 165–166.

Kennard, D. 2004. *A Shepherd's Watch. Through the Seasons with One Man and His Dogs.* London: Headline.

———. 2005. *The Dogs of Windcutter Down.* London: Headline.

Kilgour, R. 1987. "Learning and the training of farm animals." In *Farm animal Behaviour*, edited by E. O. Price, 269–284. Philadelphia: Saunders.

Kirksey, S. E. and S. Helmreich. 2010. "The emergence of multispecies ethnography." *Cultural Anthropology* 25(4): 545–576.

Kjær, H. 1927. *To votivfund fra yngre bronzealder fra Fyen og Jylland.* Aarbøger for nordisk Oldkyndighed og Historie. København.

Knight, J. 2005. "Introduction." In *Animals in Person: Cultural Perspectives on Human-Animal Intimacy*, edited by J. Knight, 1–13. New York: Berg.

Kohn, E. 2007. "How dogs dream: Amazonian natures and the politics of transspecies engagement." *Americal Ethnologist* 34(1): 3–24.

———. 2013. *How Forests Think: Towards an Anthropology Beyond the Human.* Berkeley: University of California Press.

Kristiansen, K. 1993. "Erhvervsstrategi, bebyggelse og økologi i bronsealderen - nye resultater fra Nordvest Jylland." In *Ekonomi och näringsformer i nordisk bronsålder: rapport från det 6e Nordiska bronsålderssymposiet, Nämforsen 1990*, edited by L. Forsberg and T. B. Larson, 27–34. Umeå: Umeå Universitet.

———. 1998. *Europe Before History.* New Studies in Archaeology. Cambridge: Cambridge University Press.

Kristiansen, K. and T. B. Larsson. 2005. *The Rise of Bronze Age Societies: Travels, Transmissions and Transformations.* Cambridge: Cambridge University Press.

Kristoffersen, S. 1995. "Transformation in Animal Art." *Norwegian Archaeological Review* 28: 3–15.

Kristoffersen, S. and K. Armstrong Oma. 2008. *Solens reise. Rogaland i europeisk bronsealder.* AmS Småtrykk 80. Stavanger: Arkeologisk museum i Stavanger.

Krohn, C. C., J. G. Jago and X. Boivin. 2001. "The effect of early handling on the socialisation of young calves to humans." *Applied Animal Behaviour Science* 74(2): 121–133.

Kvalø, F. 2002. "Oversjøiske reiser fra sørvest-norge til nordvest-jylland i eldre bronsealder." In *Sjøreiser og stedsidentitet. Jæren/Lista i bronsealder og eldre jernalder*, edited by L. Hedeager, 13–134. Oslo: Unipub.

Kveiborg, J. 2008. "Fårehyrder, kvægbønder eller svineavlere. En revurdering af jernalderens dyrehold." *Kuml* 59–95.

———. 2009. *Bondens dyr - husdyrhold i ældre jernalder belyst ved fund af indebrændte dyr.* Delrapport IV. Ginderup. Århus: Moesgård Museum.

Kymlicka, W. and S. Donaldson. 2011. *Zoopolis: A Political Theory of Animal Rights.* Oxford: Oxford University Press.

Lagerås, P. and M. Regnell. 1999. "Agrar förändring under sydsvensk bronsålder." In *Spiralens öga. Tjugo artiklar kring aktuell bronsåldersforskning*, edited by M. Olausson, 263–276. Riksantikvarieämbetet.

Larsson, T. B. 2002. "De döda, de "andra" och djuren." In *Bilder av bronsålder*, edited by J. Goldhahn, 91–112. Stockholm: Almqvist & Wiksell International.

Law, J. and A. Mol. 2010. "The actor-enacted: Cumbrian sheep in 2001." In *Material Agency: Towards a Non-Anthropocentric Approach*, edited by C. Knappett and L. Malafouris, 57–76. New York: Springer.

Laycock, S. W. 1999. "The animal *as* animal: A plea for open conceptuality." In *Animal others: On Ethics, Ontology and Animal Life*, edited by H. P. Steeves 271–284. New York: State University of New York Press.

Lévi-Strauss, C. 1963. *Structural Anthropology.* New York: Basic Books.

———. 1964 [1962] *Totemism.* Translated by R. Needham. London: Merlin Press.

Lia, Ø. 2001. " Det rituelle rom. En fortolkende analyse av vikingtidens graver og landskap på Kaupang." Unpublished MA Thesis, University of Oslo.

Lillehammer, G. 2007. "The past in the present" Landscape perception, archaeological heritage and marginal farmland in Jæren, South-western Norway." *Norwegian Archaeological Review*, 40: 159–178.

Løken, T. 1987. "Forsand: nå også med den første kjente bronsealderlandsby i Norge." *Frá haug ok heidni* 11(3): 236–243.

———. 1988. *Bygg fra fortiden—Forsand i Rogaland—bebyggelses-sentrum gjennom 2000 år*. AmS-Småtrykk 21. Stavanger: Arkeologisk museum i Stavanger.

———. 1989. "Rogalands bronsealderboplasser—sett i lys av områdets kulturelle kontakter." In *Rapport fra 5. Nordiske bronsealdersymposium, på Sandbjerg slott 1987*, edited by J. Paulsen, 141–147. Jysk Arkæologisk Selskabs Skrifter 24. Aarhus: Aarhus Universitetsforlag.

———. 1997. "Det forhistoriske huset i Rogaland—belyst ved flateavdekkende utgravninger." *Hus och tomt i Norden under förhistorisk tid, Bebyggelseshistorisk tidskrift* 33: 169–184.

———. 1998a. "Hustyper og sosial struktur gjennom bronsealder på Forsandmoen, Rogaland, Sørvest-Norge." In *Bronsealder i Norden—regioner og interaksjon. Foredrag ved det 7. Nordiske bronsealdersymposium i Rogaland 31. August - 3. September 1995*, edited by T. Løken, 107–121. AmS-Varia 33. Stavanger: Arkeologisk museum i Stavanger.

———. 1998b. "The longhouses of Western Norway from the Late Neolithic to the 10th century AD: Representatives of a common Scandinavian building tradition or a local development?" *NIKU Temahefte: Grindbygde hus i Vest-Norge* 30: 52–64.

———. 2003. "Utgravningsrapport for Forsandprosjektet 1980-1990. Forsand gnr. 41 og Berge gnr. 37. Forsand kommune." Innberetning til topografisk arkiv. Stavanger: Arkeologisk museum Stavanger.

Longton, T. and B. Sykes. 1997. *Training the Sheep Dog*. Ramsbury: Crowood Press.

Lorenz, K., M. Martys and A. Tipler. 1991. *Here Am I, Where Are You? The Behavior of Greylag Goose*. New York: Harcourt Brace Jovanovich.

Lovelock, J. 1992 [1991]. *Gaia: Søkelys på jordens helsetilstand*. Oslo: Cappelen.

Lund, V. 2002. "Ethics and animal welfare in organic animal husbandry—an interdisciplinary approach." Unpublished PhD thesis, Swedish University of Agricultural Sciences.

———. 2005. "The human-animal relationship in organic farming." In *The Human-Animal Relationship: Forever and a Day*, edited by F. de Jonge and R. van den Bos, 231–246. Assen: Royal van Gorcum.

Lund, V., R. Anthony and H. Röcklinsberg. 2004a. "The ethical contract as a tool in organic animal husbandry." *Journal of Agricultural and Environmental Ethics* 17 (1): 23–49.

Lund, V., S. Hemlin and J. White. 2004b. "Natural behaviour, animal rights, or making money—a study of Swedish organic farmers' view of animal issues." *Journal of Agricultural and Environmental Ethics* 17(2): 157–179.

Lund, V. and H. Röcklinsberg. 2001. "Outlining a concept of animal welfare for organic farming systems." *Journal of Agricultural and Environmental Ethics* 14 (4): 391–424.

Mandt, G. and T. Lødøen. 2005. *Bergkunst. Helleristningar i Noreg*. Oslo: Det Norske Samlaget.

Marchini, L. 2016. "Must Farm: An extraordinary tale of the everyday." *Current Archaeology* 319. https://www.archaeology.co.uk/articles/features/must-farm.htm.

Marciniak, A. 2005. *Placing Animals in the Neolithic: Social Zooarchaeology of Prehistoric Farming Communities*. London: UCL Press.

Markowitz, T. M., M. R. Dally, K. Gursky and E. O. Price. 1998. "Early handling increases lamb affinity for humans." *Animal Behaviour* 55(3): 573–587.

Marres, N. 2005. "Issues spark a public into being: A key but often forgotten point of the Lippman-Dewey debate." In *Making Things Public: Atmospheres of Democracy*, edited by B. Latour and P. Weibel, 208–217. Cambridge, MA: MIT Press.

Martin, C. F., D. Biro and T. Matsuzawa. 2011. "Chimpanzees' use of conspecific cues in matching-to-sample tasks: public information use in a fully automated testing environment." *Animal Cognition* 14(6): 893–902.

Marzluff, J. and T. Angell. 2012. *The Gifts of the Crow: How Perception Emotion and Thought Allow Smart Birds to Behave Like Humans*. Oxford: Simon & Schuster.

Mazoyer, M. and L. Roudart. 2006. *A History of World Agriculture: From the Neolithic Age to the Current Crisis*. London: Earthscan.

McFarland, S. E. and R. Hediger. 2009. "Approaching the agency of other animals." In *Animals and Agency: An Interdisciplinary Exploration*, edited by S. E. McFarland and R. Hediger, 1–20. Leiden: Brill.

McNiven, I. J. 2010. "Navigating the human-animal divide: marine mammal hunters and rituals of sensory allurement." *World Archaeology* 42(2): 215–230.

Meadows, J. R., I. Cemal, O. Karaca, E. Gootwine and J. W. Kijas. 2007. "Five ovine mitochondrial lineages identified from sheep breeds of the near East." *Genetics* 175(3): 1371–1379. https://doi.org/10.1534/genetics.106.068353.

Mech, L. D. 1999. "Alpha status, dominance and division of labor in wolf packs." *Canadian Journal of Zoology* 77: 1196–1203.

Melheim, A. L. 2004. "I mørket er alle katter grå. Om overgangen til branngravskikk i bronsealderen." In *Mellom himmel og jord. Foredrag fra et seminar om religionsarkeologi Isegran 31. januar - 2. februar 2002*, edited by L. Melheim, L. Hedeager and K. Oma, 392–421. Oslo: Oslo University, Department of Archaeology, Art History and Conservancy.

———. 2006. "Gjennom ild og vann. Graver og depoter som kilde til kosmologi i bronsealderen i Øst-Norge." In *Myter og religion i bronsealderen. Studier med utgangspunkt i helleristninger, graver og depoter i Sør-Norge og Bohuslän*, edited by C. Prescott, 13–194. Oslo: Oslo University, Department of Archeology, Art History and Conservancy.

Midgley, M. 1988. "Beasts, brutes and animals." In *What is An Animal?*, edited by T. Ingold, 35–46. London: Unwin Hyman.

———. 1996. *Utopias, Dolphins and Computers: Problems of Philosophical Plumbing*. London: Routledge.

Mikkelsen, M. 1997. "Nr 209, Legård. Arkæologiske udgravninger i Danmark 1996." Unpublished report.

Miller, D., ed. 2005. *Materiality*. Durham, NC: Duke University Press.

Milton, K. 2002. *Loving Nature: Towards An Ecology of Emotion*. London: Routledge.

———. 2005. "Anthropomorphism or egomorphism? The perception of non-human persons by human ones." In *Animals in Person: Cultural Perspectives on Human-Animal Intimacy*, edited by J. Knight, 255–271. New York: Berg.

Minahen, C. D. 1997. "Humanimals and anihumans in Gary Larson's gallery of the absurd." In *Animal Acts: Configuring the Human in Western History*, edited by J. Ham and M. Senior, 231–251. London: Routledge.

Mlekuž, D. 2007. "'Sheep are your mother': rhyta and the interspecies politics in the Neolithic of the eastern Adriatic." *Documenta Praehistorica* 34: 267–280.

———. 2012. "Agencies of assemblages: flocks, herds, households." Paper presented at the European Association of Archaeologists Meeting, 2012, Helsinki, 31.08.2012.

———. 2013. "The birth of the herd." *Society & Animals* 21(2): 150–161.

Møllerop, O. 1968. "Helleristningene på Dysjaland." *Fra Haug og Heidni* 2: 144–146.

Morris, B. 2000 [1998]. *The Power of Animals: An Ethnography*. New York: Berg.
———. 2000. *Animals and Ancestors: An Ethnography*. New York: Berg.
Mullin, M. H. 1999. "Mirrors and windows: Sociocultural studies of human-animal relationship." *Annual Review of Anthropology* 28: 201–224.
Myhre, B. 1979. "Høvdinggraver fra eldre bronsealder på Rege i Sola." *Frå haug ok Heidni* 3: 230.
———. 1995. "Skipsgravene på Karmøy. Eit forskningsprosjekt på Avaldsnes." In *Rapport frå Arkeologisk museum i Stavanger oktober 1994*, edited by A. Lillehammer, 79–90. Stavanger: Arkeologisk museum i Stavanger.
Myhre, B. and I. Øye. 2002. *Norges landbrukshistorie I: 4000 f.Kr - 1350 e.Kr. Jorda blir levevei*. Oslo: Det Norske Samlaget.
Nadasdy, P. 2007. "The gift in the animal: The ontology of hunting and human-animal sociality." *American Ethnologist* 34(1): 25–43.
Nagasawa, M., T. Kikusui, T. Onaka and M. Ohta 2009. "Dog's gaze at its owner increases owner's urinary oxytocin during social interaction." *Hormones and Behaviour* 55: 434–441.
Nagel, T. 1974. "What is it like to be a bat?" *The Philosophical Review* 83(4): 435–450.
———. 1986. *The View From Nowhere*. Oxford: Oxford University Press.
Nielsen, J. N. 2002. "Flammernes bytte." *Skalk* (6): 5–10.
Nordenborg Myhre, L. 1998. *Historier fra en annen virkelighet. Fortellinger om bronsealderen ved Karmsundet*. AmS-Småtrykk 46. Stavanger: Arkeologisk museum i Stavanger.
———. 2004. *Trialectic Archaeology: Monuments and Space in Southwest Norway 1700-500 BC*. AmS-Skrifter 18. Stavanger: Arkeologisk museum i Stavanger.
Noske, B. 1997. *Beyond Boundaries: Humans And Animals*. Montreal: Black Rose Books.
Nyegaard, G. 1996. "Faunalevn fra bronzealder: En zooarkæologisk undersøgelse af sydskandinaviske boplatsfund." Unpublished PhD thesis, University of Copenhagen.
O'Connor, T. P. 1997. "Working at relationships: Another look at animal domestication." *Antiquity* 71: 149–156.
———. 2001. "Economic prehistory or environmental archaeology? On gaining a sense of identity." In *Environmental Archaeology: Meaning and Purpose*, edited by U. Albarella, 17–27. London: Kluwer Academic Publishers.
———. 2013. *Animals As Neighbors: The Past and Present of Commensal Animals*. East Lansing: Michigan State University Press.
Olausson, M. 1999. "Herding and stalling in Bronze Age Sweden." In *Settlement and Landscape*, edited by C. Fabeck and J. Ringtved, 319–328. Århus: Jutland Archaeological Society.
Olmert, M. 2009. *Made for Each Other: The Biology of the Human-Animal Bond*. Philadelphia, PA: Da Capo Press.
Olsen, A.-L. H. 2007. "Gård i flammer." *Skalk* 5: 3–10.
Olsen, B. 2003. "Material culture after text: re-membering things." *Norwegian Archaeological Review* 36(2): 87–104.
Oma, K. 2000. "Hesten i nordisk jernalder. Ei kontekstuell analyse av den symbolske sfære kontra den materialle røynda." Unpublished MA thesis, Faculty of Historical Philosophy, University of Oslo.
———. 2001. "Hesten i jernalderen - i brytningspunktet mellom "seige strukturar" og endring i den materielle kulturen." *Primitive Tider* 4: 38–50.
———. 2004. "Hesten og det heilage. Materialiseringa av eit symbol." In *Mellom himmel og jord. Foredrag fra et seminar om religionsarkeologi, Isegran 31. januar - 2. februar 2002*, edited by L. Melheim, L. Hedeager and K. Oma, 68–81. Oslo: University of Oslo.

Orton, D. 2010. "Both subject and object: Herding, inalienability and sentient property in prehistory." *World Archaeology* 42(2): 188–200.
Overton, N. and Y. Hamilakis. 2013. "A manifesto for a social zooarchaeology: Swans and other beings in the Mesolithic." *Archaeological Dialogues* 20(2): 111–136.
Padel, R. 2004. *The Soho Leopard*. London: Chatto & Windus.
Pappa, M., P. Halstead, K. Kotsakis and D. Urem-Kotsou. 2004. "Evidence for large-scale feasting at late Neolithich Makriyalos, N Greece." In *Food, Cuisine and Society in Prehistoric Greece*, edited by P. Halstead, J. C. Barrett, 16–44. Sheffield Studies in Aegean Archaeology 5. Oxford: Oxbow Books.
Patrik, L. E. 2000. "Is there an archaeological record?" In *Interpretive Archaeology: A reader*, edited by J. Thomas, 118–144. London: Leicester University Press.
Payne, S. 1973. "Kill-off patterns in sheep and goats: The mandibles from Asvan Kale." *Anatolian Studies* 23: 281–303.
Pedersen, G. M. 2013. "Arkeologisk undersøkelse av hus fra eldre bronsealder og gårdsanlegg fra yngre romertid/folkevandringstid. Lunde gnr 4, bnr. 1, 13, Stavanger kommune, Rogaland." *Oppdragsrapport* 2013/12. Stavanger: Arkeologisk museum, Universitetet i Stavanger.
Pepperberg, I. M. 2000. *The Alex Studies: Cognitive and Communicative Abilities of Grey Parrots*. Cambridge, MA: Harvard University Press.
Petersson, M. 2006. *Djurhållning och betesdrift: djur, människor och landskap i västra Östergötland under yngre bronsålder och äldre järnålder*. Linköping: Riksantikvarieämbetet, Uppsala Universitet.
Philo, C. and C. Wilbert. 2000. *Animal Spaces, Beastly Places: New Geographies of Human-Animal Relations*. London: Routledge.
Pilley, J. W. and A. K. Reid. 2011. "Border collie comprehends object names as verbal referents." *Behavioural processes* 86(2): 184–195.
Pilskog, F. 1998a. "Arkeologisk utgraving på Sandved i Sandnes kommune—smakebitar frå undersøkinga sommaren 1998." *Frá haug ok heidni* 4: 13–16.
———. 1998b. "Innberetning til topografisk arkiv. Rapport om den arkeologiske undersøkinga på Sandved, Skeiane 40/246 og Brueland 41/22, Sandnes k., Rogaland fylke." Unpublished report. Stavanger: Arkeologisk museum i Stavanger.
Porteous, D. J. 1985. "Smellscape." *Progress in Human Geography* 9(3): 356–378.
Prescott, C. 1991a. *Kulturhistoriske undersøkelser i Skrivarhelleren*. Bergen: Historisk museum, Universitetet i Bergen.
———. 1991b. "Late Neolithic and Bronze Age developments on the periphery of Southern Scandinavia." *Norwegian Archaeological Review* 24(1): 35–48.
———. 1995. "Aspects of early pastoralism in Sogn, Norway." *Acta Archaeologica* 66: 163–190.
———. 1996. "Was there really a Neolithic in Norway?" *Antiquity* 70(1): 77–87.
———. 1999. "Long-term patterns of non-agrarian exploitation in southern Norwegian highlands." In *Settlement and Landscape*, edited by C. Fabeck and J. Ringtved, 213–223. Århus: Jutland Archaeological Society.
———. 2012. "The origin of a Bronze Age in Norway: Structure, regional process and localized history." In *Local Societies in Bronze Age Northern Europe*, edited by N. Anfinset and M. Wrigglesworth, 215–231. Sheffield: Equinox Publishing.
Prescott, C. and A. L. Melheim. 2017. "Textiles from the peripheries? Upland evidence from Norway." In *New perspectives on the Bronze Age: Proceedings of the 13th Nordic Bronze Age Sympisium held in Gothenburg 9th to 13th June 2015*, edited by S. Bergerbrant and A. Wessmann, 313–325. Oxford: Archaeopress.
Price, C. 2009. *Border Collie: An owner's guide*. London: Harper Collins.
———. 2011. "Social security." *Your dog*, 12–15.

Price, E. O. 1984. "Behavioural aspects of animal domestication." *Quarterly Review of Biology* 59(1): 1–32.

Prøsch-Danielsen, L. and A. Simonsen 2000a. *The Deforestation Patterns and the Establishment of the Coastal Heathland of South-Western Norway*. AmS-Skrifter 15. Stavanger: Arkeologisk museum i Stavanger.

———. 2000b. "Palaeoecological investigations towards the reconstruction of the history of forest clearances and coastal heathlands in south-western Norway." Vegetation History and Archaeobotany 9: 189–204.

Pryor, F. 2006 [1998]. *Farmers in Prehistoric Britain*. Stroud: Tempus.

Randsborg, K. 2011. *Bronze Age Textiles: Men, Women and Wealth*. London: Bristol Classical.

Rasmussen, J. L. and D. W. Rajecki. 1995. "Differences and similarities in human perception of the thinking and feeling of a dog and a boy." *Society and Animals* 3(2): 117–137.

Rasmussen, M. 1999. "Livestock without bones: The long house as contributor to the interpretation of livestock management in the Southern Scandinavian Early Bronze Age." In *Settlement and Landscape*, edited by C. Fabeck and J. Ringtved, 281–290. Århus: Jutland Archaeological Society.

Rasmussen, M. and C. Adamsen. 1993a. "Bebyggelsen." In *Da klinger i muld*, edited by J. Jensen, 136–141. Århus: Jysk Arkæologisk Selskab.

———. 1993b. "[The Bronze Age] Settlement." In *Digging into the Past. 25 Years of Archaeology in Denmark*, edited by S. B. S. Hvass, 136–141. Aarhus: Royal Society of Northern Antiquaries / Jutland Archaeological Society.

Rast-Eicher, A. and L. B. Jörgensen. 2013. "Sheep wool in Bronze Age and Iron Age Europe." *Journal of Archaeological Science* 40(2): 1224–1241.

Rebanks, J. 2016. *The Shepherd's Life: A Tale of the Lake District*. London: Penguin Press.

Regan, T. 1988. *The Case For Animal Rights*. London, Routledge.

Reiersen, K. 2015. "Arkeologiske undersøkelser av hus og øvrige bosetningsspor fra eldre bronsealder og førromersk jernalder, ID177381. Nordre Tengesdal gnr. 81, bnr. 4, Bjerkreim kommune, Rogaland." *Oppdragsrapport* 2015/25. Stavanger: Arkeologisk museum, Universitetet i Stavanger.

Reitz, E. J. and E. S. Wing. 1999. *Zooarchaeology*. Cambridge Manuals in Archaeology. Cambridge: Cambridge University Press.

Renfrew, C. 2004. "Towards a theory of material engagement." In *Rethinking Materiality: The Engagement of Mind with the Material World*, edited by E. DeMarrais, C. Gosden and C. Renfrew, 23–32. Cambridge: MacDonald Institute.

Rindos, D. 1984. *The Origins of Agriculture: An Evolutionary Perspective*. New York: Academic Press.

Rizzolatti, G., M. Fabbri-Destro and L. Cattaneo. 2009. "Mirror neurons and their clinical relevance." *Nature Clinical Practice Neurology* 5(1): 24–34.

Rousing, T. and F. Wemelsfelder. 2006. "Qualitative assessment of social behaviour of dairy cows housed in loose housing systems." *Applied Animal Behaviour Science* 101: 40–53.

Roymans, N. 1999. "Man, cattle and the supernatural in the Northwest European plain." In *Settlement and Landscape*, edited by C. Fabeck and J. Ringtved, 291–300. Århus: Jutland Archaeological Society.

Russell, N. 2012. *Social Zooarchaeology: Humans and Animals in Prehistory*. Cambridge: Cambridge University Press.

Ryder, M. L. 1981. "A survey of European primitive breeds pf sheep." *Annales de Génétique et de Sélection Animale* 13(4): 381–418.

———. 1983. *Sheep and Man*. London: Duckworth.

Ryder, R. D. 1971. "Experiments on animals." In *Animals, Men and Morals: An Enquiry into the Maltreatment of Non-Humans*, edited by S. Godlovitch, R. Godlovitch and J. Harris, 41–82. London: Victor Gollanz.

Særheim, I. 2011. "Hafrsfjord og nokre andre stadnamn kring fjorden på Nord-Jæren." In *Tverrvitenskapelige perspektiver 2*, 5–14. Stavanger: Arkeologisk museum, Universitetet i Stavanger.

———. 2015. *Jærstrendene. Namn og stader.* AmS-Varia 57. Stavanger: Arkeologisk museum, Universitetet i Stavanger.

Sæther, K. E. 2007. "From settlement to way of life: A socio-economic analysis of Rogaland's Early Bronze Age settlements." Unpublished MA thesis, University of Oslo.

Saint-Exupéry, A. D. 1995 [1943]. *The Little Prince.* London: Wordsworth Classics.

Sanders, C. R. 1993. "Understanding dogs: caretakers' attribution of mindedness in canine-human relationships." *Journal of Contemporary Etnography* 22(2): 205–226.

Schwabe, C. W. 1984. "Drinking cow's milk: The most intense man-animal bond." In *The Pet Connection: Its Influence On Our Health and Quality of Life*, edited by R. K. Anderson, B. L. Hart and L. A. Hart, 50–57. Minneapolis: Center to Study Human–Animal Relationships and Environments, University of Minnesota.

Scott, J. P. 1962. "Introduction to animal behaviour." In *The Behaviour of Domestic Animals*, edited by E. S. E. Hafez, 3–20. London: Ballière, Tindall & Cox.

Scrimgeour, D. 2002. *Talking Sheepdogs: Training Your Working Border Collie.* Preston: Good Life Press.

Seabrook, M. F. 1972. "A study to determine the influence of a herdsman's personality on milk yield." *Journal of Agricultural Labour Science* 1: 44–59.

———. 2001. "The effect of the operational environment and operating protocols on the attitudes of and behaviour of employed stockpersons." Human–Animal relationship: stockmanship and housing in organic livestock systems. Proceedings of the third NAHWOA workshop, Clermont Ferrand, 21–24 October 2000, Clermont Ferrand.

Seed, A., N. Emery and N. Clayton. 2009. "Intelligence in corvids and apes: a case of convergent evolution?" *Ethology* 115 (5):401–420.

Serjeantson, D. 1996. "The animal bones." In *Runnymede Bridge research excavations. Refuse and disposal at area 16 east, Runnymede*, edited by S. Needham and T. Spence, 194-223. London: British Museum Press.

———. 2007. "Intensification of animal husbandry in the Late Bronze Age? The contribution of sheep and pigs." In *The Earlier Iron Age in Britain and the Near Continent*, edited by C. Haselgrove and R. Pope, 80–93. Oxford: Oxbow Books.

Serpell, J. A. 1996 [1986]. *In the Company of Animals. A Study of Human-Animal Relationships.* Cambridge: Cambridge University Press.

Shanklin, E. 1985. "Sustenance and symbol: anthropological studies of domesticated animals." *Annual Review of Anthropology* 14: 375–403.

Shapiro, K. J. 2008. *Human-Animal Studies: Growing the Field, Applying the Field.* Ann Arbor, MI: Animals & Society Institute.

Shapiro, K. J. and M. DeMello. 2010. "The state of Human–Animal Studies." *Society & Animals* 18(3): 307–318.

Sheets-Johnstone, M. 1999. "Phenomenology and agency: Methodological and theoretical issues in Strawson's 'The Self'." *Journal of Consciousness Studies* 6: 48–69.

Sherratt, A. 1981. "Plough and pastoralism: Aspects of secondary products revolution." In *Patterns of the Past: Studies in Honour of David Clarke*, edited by I. Hodder, G. Isaac and N. Hammond, 261–305. Cambridge: Cambridge University Press.

Singer, P. 1975. *Animal Liberation.* New York: Random House.

Sofaer Derevenski, J. 2000. "Material culture shock: Confronting expectations in the material culture of children." In *Children and Material Culture*, edited by J. Sofaer Derevenski, 3–16. London: Routledge.

Solbakken, A. C. 2010. "Atferdskonsultasjoner over en tiårsperiode." *Hundesport. Tidsskrift for norsk kennelklubb* 111(9–10): 50–51.

Soltvedt, E.-C. 2017. "Up and down?" Paper presented at the Nordic Archaeobotanic group meeting, Helsinki.

———. In prep. "Cereal cultivation in southwest Norway: Boom and bust in the Late Neolithic and Early Bronze Age." Unpublished manuscript.

Soltvedt, E.-C., T. Løken, L. Prøsch-Danielsen, R. Børsheim and K. Oma. 2007. *Bøndene på Kvålehodlene. Boplass-, jordbruks-, og landskapsutvikling gjennom 6000 år på Jæren, SV Norge*. Vol. 47. AmS-Varia 47. Stavanger: Arkeologisk Museum i Stavanger.

Sørskog, O. D. 2007. "Helleristningene i et maktperspektiv. Et kontekstuelt studie av ristningane på Jæren." Unpublished MA Thesis, Archaeology, University of Bergen.

Startin, W. 1978. "Linear pottery culture houses: Reconstruction and manpower." *Proceedings of the Prehistoric Society* 44: 143–159.

Steeves, H. P. 1999. "Introduction." In *Animal Others: On Ethics, Ontology and Animal Life*, edited by H. P. Steeves, 1–14. Albany: State University of New York Press.

Stig Sørensen, M.-L. 1997. "Reading Dress: the Construction of Social Categories in Bronze Age Europe." *Journal of European Archaeology* 5: 93–114.

———. 2000. *Gender Archaeology*. Oxford: Blackwell.

Strathern, M. 1988. *The Gender of the Gift: Problems with Women and Problems with Society in Melanesia*. Berkely: University of California Press.

Streiffert, J. 2001. *På gården. Rumslig organisation inom bosättingsytor och byggnader under bronsålder och äldre järnålder*. Vol. 35, *Riksantikvarieämbetet Arkeologiska Undersökningar Skrifter; Gotarc. Serie C. Arkeologiska Skrifter*. Stockholm: Riksantikvarieämbetet & Göteborgs universitet Institutionen för arkeologi.

Swabe, J. 2005. "Loved to death? Veterinary visions of pet-keeping in modern Dutch society." In *Animals in Person: Cultural Perspectives on Human-Animal Intimacy*, edited by J. Knight, 101–118. New York: Berg.

Sykes, B. 1999. *Understanding Border Collies*. Ramsbury, Wiltshire: Crowood Press.

Syvertsen, K. I. J. 2003. "Ristninger i graver – graver med ristninger. Om ristningers mening i gravminner og gravritualer. En analyse av materiale fra Rogaland." Unpublished MA thesis, Department of Archaeology, University of Bergen.

———. 2005. "Rogalands ristninger i graver som transformerende og stabiliserende faktorer i tilværelsen." In *Mellan sten och järn. Rapport från det 9:e nordiska bronsålderssymposiet, Göteborg 2003-10-09/12*, edited by J. Goldhahn, 503–520. Göteborg: Gotarc Serie C, Arkeologiske skrifter.

Tapio, M., N. Marzanov, M. Ozerov, M. Cinkulov, G. Gonzarenko, T. Kiselyova, M. Murawski, H. Viinalass and J. Kantanen. 2006. "Sheep mitochondrial DNA variation in European, Caucasian and Central Asian areas." *Molecular Biology and Evolution* 23(9): 1776–1783.

Tesch, S. 1992. "House, farm and village in the Köpinge area from Early Neolithic to the Early Middle Ages." In *The Archaeology of the Cultural Landscape: Field Work and Research in a South Swedish Rural Region*, edited by L. Larson, J. Callmer and B. Stjärnquist, 283–344. Stockholm: Almqvist & Wiksell.

———. 1993. "Houses, farmsteads and long term change: A regional study of prehistoric settlements in the Köpinge area, in Scania, Southern Sweden." Unpublished PhD thesis, Uppsala University.

Theodossopoulos, D. 2005. "Care, order and usefulness: The context of the human-animal relationship in a Greek island community." In *Animals in Person: Cultural Perspectives on Human-Animal Intimacy*, edited by J. Knight, 15–35. New York: Berg.

Thomas, J. 2004. *Archaeology and Modernity*. London: Routledge.

Thomas, K. 1984 [1983]. *Man and the Natural World: Changing Attitudes in England 1500–1800*. London: Penguin Books.

Thomas, K. D. 1996. "Zooarchaeology: Past, present and future." In *Zooarchaeology*, 1–4. London: Routledge.

Thompson, K. 2011. "Theorising rider-horse relations: An ethnographic illustration of the centaur metaphor in the Spanish bullfight." In *Theorizing animals: rethinking humanimal relations*, edited by N. Taylor and T. Signal, 221–253. Leiden: Brill.

Thorsen, L. E. 2001. *Hund! Fornuft og følelser*. Oslo: Pax Forlag.

Trigger, B. 2006. *A History of Archaeological Thought*. Cambridge: Cambridge University Press.

Tsing, A. 2012. "Unruly edges: Mushrooms as companion species. For Donna Haraway." *Environment and Society* 1: 141–154.

Ullén, I. 1994. "The power of case studies: Interpretation of a Late-Bronze-Age settlement in central Sweden." *Journal of European Archaeology* 2(2): 249–262.

———. 1995a. "Det goda exemplets makt. Närstudie av en bronsåldersbosätning I Uppland." In *Samhällstruktur och förändring under bronsåldern. Rapport från ett bronsålderseminarium på Norrköpings stadsmuseum*, edited by M. Larsson, A. Toll, 68–75. Norrköping: Riksantikvarieämbetet.

———. 1995b. "Hushållet i Apalle under bronsåldern." *Populär Arkeologi* 13(3): 31–33.

———. 1996. "Horse and dog in the Swedish Bronze Age: A close-up study of the relation of horse and dog to man in the Bronze Age settlement of Apalle." *Archäologisches Korrespondenzblatt* 26: 145–166.

———. 1997. "Lager och hus." In *Bronsåldersboplatsen vid Apalle i Uppland. Arkeologi på väg - undersökningar för E18*, edited by I. Ullén, 22–75. Uppsala: Riksantikvarieämbetet.

van Post, L. 1919. "Ett par offerdammar från Skånes bronsålder." *Rig* 2: 160–176.

Vevatne, K. 1996. "Ristningar i Etne - ein analyse av tid og rom." Unpublished MA thesis, Archaeology, University of Bergen.

Viveiros de Castro, E. 1998. "Cosmological deixis and Amerindian perspectivism." *The Journal of the Royal Anthropological Institute* 4(3): 469–488.

———. 2004. "Exchanging perspectives: The transformation of objects into subjects in Amerindian ontologies." *Common Knowledge* 10(3): 463–484.

Waterbolk, H. T. 1975. "Evidence of cattle stalling in excavated pre- and protohistoric houses." In *Archaeozoological studies*, edited by A. T. Clason, 383–394. Amsterdam: North-Holland Publishing Company.

Wemelsfelder, F. 2001. "The inside and outside aspects of consciousness: Complementary approaches to the study of animal emotion." *Animal Welfare* 10: 129–139.

———. 2012. "A science of friendly pigs... Carving out a conceptual space for addressing animals as sentient beings." In *Crossing boundaries: Investigating human-animal relationships*, edited by L. Birke and J. Hockenhull, 223–249. Leiden: Brill.

Wemelsfelder, F., T.E.A. Hunter, M. T. Mendl and A. B. Lawrence. 2001. "Assessing the 'whole animal': A free choice profiling approach." *Animal Behaviour* 62: 209–220.

Wengrow, D. 2006. *The Archaeology of Early Egypt: Social Transformations in North East Africa, 10,000 to 2650 BC*. Cambridge World Archaeology. Cambridge: Cambridge University Press.

Westling, L. 2014. "The zoosemiotics of sheep herding with dogs." In *The Semiotics of Animal Representation*, edited by M. Tønnesen and K. Tüür, 33–52. Amsterdam: Rodopi.

Westling, S. and A. Overland. 2012. "Naturvitenskaplige undersøkelser på Kvia-Motland. Kvia gnr. 19/31, og Motland gnr. 20/3, Hå k., Rogaland." *Oppdragsrapport* 2012/3. Universitetet i Stavanger, Arkeologisk museum.

Whittle, A. 2003. *The Archaeology of People. Dimensions of Neolithic Life*. London: Routledge.

Wilkie, R. M. 2010. "Livestock/deadstock: Working with farm animals from birth to slaughter." In *Working with Farm Animals from Birth to Slaughter*, 147–172. Philadelphia, PA: Temple University Press.

Willerslev, R. 2007. *Soul Hunters: Hunting, Animism and Personhood among the Siberian Yukaghirs*. Berkeley: University of California Press.

———. 2012. *On the Run in Siberia*. Minneapolis: University of Minnesota Press.

Wolfe, C. 2009. *What is Posthumanism?* Vol. 8, *Posthumanities*. Minneapolis: University of Minnesota Press.

Wrigglesworth, M. 2011. "Finding your place: Rock art and local identity in West Norway. A study of Bronze Age rock art in Hardanger and Sunnhordland." Unpublished PhD thesis, University of Bergen.

Young, R. 2003. *The Secret Life of Cows*. Preston: Farming Books and Videos Ltd.

Zeder, M. A. 2008. "Domestication and early agriculture in the Mediterranean Basin: Origins, diffusion and impact." *Proceedings of the National Academy of Sciences of the United States of America* 105(33): 11597–11604. https://doi.org/10.1073/pnas.0801317105.

Zeder, M. A., E. Emshwiller, B. D. Smith and D. G. Bradley. 2006. "Documenting domestication: The intersection of genetics and archaeology." *TRENDS in Genetics* 22(3): 139–155.

Zeuner, F. E. 1963. *A History of Domesticated Animals*. London: Hutchkinson.

Zimmerman, H. W. 1999a. "Favourable conditions for cattle farming, one reason for the Anglo-Saxon migration over the North Sea? About the byre's evolution in the area south and east of the North Sea and England." In *In Discussion with the Past: Archaeological Studies Presented to W.A. van Es*, edited by H. Sarfatij, W.J.H. Verwers and P. J. Woltering, 129–144. Amersfoort: Zwolle.

———. 1999b. "Why was cattle-stalling introduced in prehistory? The significance of byre and stable and of outwintering." In *Landscape and Settlement*, edited by C. Fabeck and J. Ringtved, 301–318. Århus: Jutland Archaeological Society.

Index

Numbers in *italic* denote pages with figures, numbers in **bold** denote tables.

www.ingramcontent.com/pod-product-compliance
Ingram Content Group UK Ltd.
Pitfield, Milton Keynes, MK11 3LW, UK
UKHW061401220626
472208UK00010B/29

* 9 7 8 1 7 8 1 7 9 2 5 1 3 *